Olfa Hajji

Agricultural drainage and sanitation course

Olfa Hajji

Agricultural drainage and sanitation course

ScienciaScripts

Imprint

Any brand names and product names mentioned in this book are subject to trademark, brand or patent protection and are trademarks or registered trademarks of their respective holders. The use of brand names, product names, common names, trade names, product descriptions etc. even without a particular marking in this work is in no way to be construed to mean that such names may be regarded as unrestricted in respect of trademark and brand protection legislation and could thus be used by anyone.

Cover image: www.ingimage.com

This book is a translation from the original published under ISBN 978-620-6-72904-4.

Publisher:
Sciencia Scripts
is a trademark of
Dodo Books Indian Ocean Ltd. and OmniScriptum S.R.L publishing group

120 High Road, East Finchley, London, N2 9ED, United Kingdom
Str. Armeneasca 28/1, office 1, Chisinau MD-2012, Republic of Moldova, Europe
Managing Directors: Ieva Konstantinova, Victoria Ursu
info@omniscriptum.com

Printed at: see last page
ISBN: 978-620-8-50646-9

Table of contents

Foreword

In agriculture, forestry and sometimes urban planning, drainage is the process of artificially encouraging the evacuation of excess water.

Drainage has existed since prehistoric times, with ancient traces found on every continent. It has generally contributed to significant improvements in productivity.

Since it was invented by the Romans, the art of drainage has never ceased to be perfected, and has become indispensable for agricultural development. Tunisia, where many arable lands are subject to intense problems such as rising water tables and permanent water stagnation, is trying to remedy these problems by relying on drainage techniques (surface or underground) to create a favorable environment for plant development and improve the profitability of farmland.

This book covers everything a student of hydraulics needs to know (the concept of drainage, agricultural drainage, installation of drainage networks, etc.).

We've tried to avoid going into too much detail in this discipline, which touches on both the agricultural and urban planning sectors, so as to make it easy to pass on information to the student.

The book is structured into five chapters as follows:

Chapter I *provides **general information on drainage,** including the definition and purpose of drainage and the different types of drainage.*

Chapter II *is dedicated to **the general description of a Drainage System** and to the main components.*

Chapter III *presents the **principle and methods of sanitation**.*

Chapter IV *describes the **main methods for sizing** an agricultural **drainage system.***

Chapter V *describes **drain clogging phenomena and the filtering materials** used to control drain silting.*

*The aim of **Chapter VI** is to present **the various problems associated with the malfunctioning of underground drainage systems.***

Chapter VII *presents the **various machines that can be used to install underground drainage pipes**.*

Chapter VIII *presents the **layout principle and the modelling approaches for the drainage system.***

Chapter I: General information

I.1. Introduction

In Tunisia, many perimeters are subject to intense problems such as rising water tables and *flooding*, the consequences of which are soil degradation and reduced productivity. In Tunisia, as elsewhere, there are two land use problems linked to natural or artificial conditions. These conditions present constraints for crop development and lower yields. They are particularly related to water and soil.

For drainage, these conditions are mainly :

- *Water stagnation and soil saturation* on the one hand, and *water salinity on* the other.

These two parameters cause asphyxiation and toxication of plants, especially when they are present for a long period, which exceeds the tolerance of the crop.

- Asphyxia: An asphyxiated plant is one that no longer receives sufficient oxygen, either because the soil is too compact or too humid
- Toxication: excess salt

We will therefore focus on the impact of drainage systems on agricultural land reclamation, and the basis for their calculation.

I.2 General information on drainage

Drainage is a hydro-agricultural management technique designed to reduce or eliminate excess water affected plots.

- The term **"agricultural drainage"** refers to all operations designed to *remove excess water* from excessively wet soils in order to improve tillage, access to plots and crop growth (figure 1).

- Drainage is therefore used to evacuate excess water in the soil that is harmful and/or water that is laden with salts and toxic to plants.

- Agricultural drainage is a major land improvement operation designed to remove excess water from the soil by laying underground pipes. Its implementation in hydromorphic soils ensures better use of farmland, by regularizing and securing production and improving working conditions and access to the field.

- "The term drainage refers to any natural or artificial operation contributing to the collection of water of any kind and from any origin to send it away from a certain area". Drainage can take the form *of ditches* or *drains*.

In hydromorphic soils, drainage often prevents waterlogging of the surface horizons during the wet season (winter and spring in particular) by ensuring faster drainage of the surface horizons and a temporary or permanent lowering of the water table in the soil

The essential aim of drainage is to tackle the causes of moisture and transform the soil into an active living environment for plant roots.

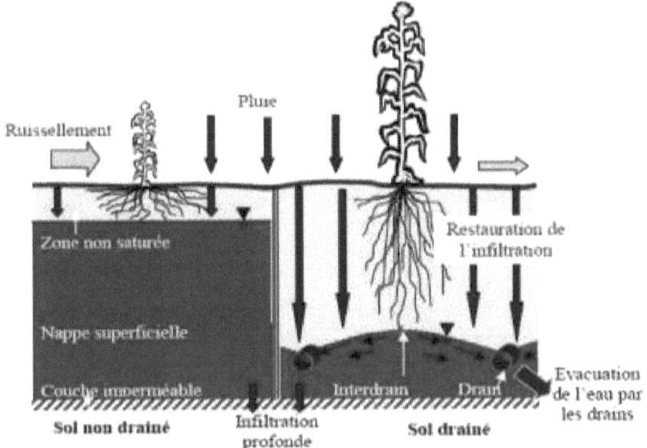

Figure 1: Concept of the water table with and without drainage (Guyomard, 2009).

Hydromorphy is a soil quality. A soil is said to be hydromorphic when it shows physical signs of regular water saturation.

Microbial life is then "drowned", and the presence of water also has physico-chemical consequences. In clay soils, hydromorphy is easy to spot.

Agricultural drainage also appears to be a cause of pollution of the natural environment, by discharging water laden with nitrates or phytosanitary treatment products.

On the other hand, the drainage of plots of land, which aims to extract excess water from the soil, is generally accompanied by an agricultural drainage network, consisting of one or more outfalls, whose role is to facilitate the transfer of drainage water downstream. The role of this network of outfalls can be better assessed and controlled, as flows follow known paths, often built for the sole purpose of channelling these flows.

Among the constraints to be taken into account when developing an outfall network (ditches, streams), the depth at which buried collectors emerge is an aggravating factor in the formation of certain floods. In fact, when water rises the most, the transfer capacity of ditches is increased by their deepening, and downstream may be subject to greater inflows.

Drainage and rural sanitation involves 3 phases:

- Capture or collect excess water (plot drainage).

- Routing through a network of collectors or ditches.
- Return to natural river system (outlet).

The aim of agricultural drainage is to prevent water stagnation during the wet season and to lower the piezometric levels of surface water tables (lowering the water table). It allows water to flow either on the surface, by installing structures or adapting the topography of the land, or underground, by building buried networks that join open ditches. The network is built on the scale of the plot or group of plots: water is collected in drains (perforated pipes) and collectors (perforated or non-perforated pipes) before being evacuated to the natural outlet (river) or in ditches.

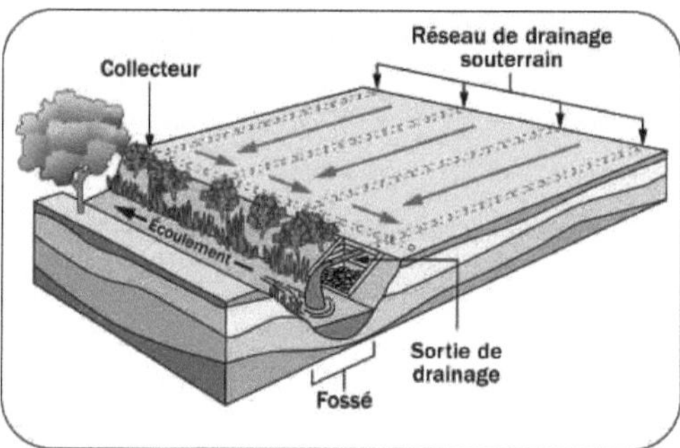

Figure 2: Simplified diagram of a drainage network

An "intense drainage season", from mid-December to the end of February, with rapid runoff response to each rainy episode.

Without a doubt, the positive impacts of drainage on soils can be summed up as a strong improvement in agricultural potential: changes in the functional characteristics of the soil (aeration, thermal regime, biological activity); improved soil structure, infiltration capacity and water circulation; reduced operating constraints (access to fields, crop diversification) and reduced risk of salinization to increase crop yields (Table 1).

Table 1: Effect of subsurface agricultural drainage on yields

of five cultures (Colwell, 1978)

Culture		Maïs-grain	Soja	Blé	Avoine	Foin
Rendement moyen avant drainage	tonne/ha	4.14	1.96	1.77	1.60	4.10
Rendement moyen après drainage	tonne/ha	5.58	2.59	2.61	2.35	5.20
Augmentation de rendement	tonne/ha	1.44	0.63	0.84	0.75	1.10
	%	34.8	32.1	47.5	46.9	26.8

To achieve these objectives, it is necessary to know the data relating to the plant, the soil and the water. These data can be supplied in the form of preliminary studies:

- Soil survey
- Hydrological study
- Agronomic study
- Topographic survey
- Economic study

I.3. Preliminary studies

I.3.1. Pedological studies

Soil surveys can be used by a drainage system designer to determine the physical characteristics of the soil (structure, texture, depth, permeability, porosity, etc.). All these characteristics can be presented in the form of a pedological map at a legible scale (1/5000 for heterogeneous soils, 1/50000 for homogeneous soils)

These soil maps can be used to delimit areas with similar characteristics, known as soil zoning. This map facilitates the choice of drainage system.

I.3.2. Hydrological study

The hydrological study will be used by a drainage system designer to determine :

- The critical rain intensity that causes water stagnation or soil saturation.
- Runoff or flood flow in waterways that cross areas exposed to saturation (e.g. plains).

Low-intensity, long-duration rainfall causing soil saturation (critical intensity).

NB: for drainage, we generally take a critical rainfall (e.g. 10 to 15 mm/d) lasting 3 successive days (return period: 1 to 2 years) determined from the IDF curve or the MONTANA formula:

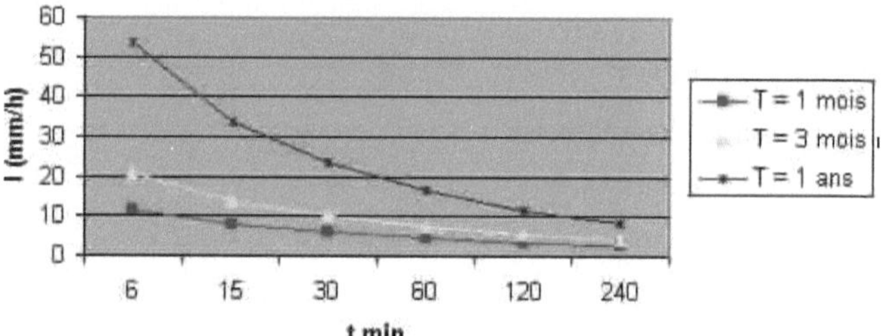

* **Formule de Montana:**

$$i_{(T)} = a_{(T)} \times t^{-b_{(T)}}$$

$i_{(T)}$: intensité en mm/h

T : temps en heures et pris égal au temps de concentration

a et b : coefficients dépendant de la fréquence de la pluie.

Généralement, **a = 7.446** et **b = 0.493**

I.3.3. Agronomic study

The agronomic study will be used by a drainage system designer to determine :

- Physiological characteristics ;
- Plant rooting depth;
- Plant tolerance to asphyxiation;
- Plant tolerance to toxication.

NB *Rooting depth for annual crops (wheat, barley, oats, sorghum, cereals) P= 70 cm to 1m.*

Rooting depth for tree crops P = 1.5 m.

* Crop tolerance to asphyxiation (to avoid yield loss):

** annual crops: 3 days and max 14 days*

** arboriculture: 7 days max 21 days to 1 month*

Salinity standards for drinking water in Tunisia (1g/l in the north and 2g/l in the south), irrigation water (1 to 4g/l).

Soil electrical conductivity Ece = salinity and Ecw : Electrical conductivity of irrigation water

Hydraulic conductivity = permeability

1mm = 1l/m² = 10 m³/ha

I.3.4. Topographical study

For a drainage system designer, a topographic survey is used to :

- Locate the drainage network outlet.
- Trace the drainage network.

This topographical study must be carried out on *a plan of the land parcel side* in which there is CN, the coasts, the land parcel units.

This side plan should be drawn to a legible scale, showing any peculiarities (bumps and depressions).

This side plan must be drawn up by a topographer familiar with drainage.

The scale required for this plan is generally 1/10000 for areas that do not present any singularities (dyke, wadi passing through the plain); and 1/5000 or 1/2000, for areas that do present singularities.

Longitudinal and cross-sectional profiles are required for setting the drainage network.

I.3.5. study

The economic study is used by a drainage system planner to choose the drainage system best suited to the money available (the most economical variant).

I.4. Experimental determination of hydrodynamic parameters

The two hydrodynamic soil parameters required for drainage are :

- Permeability **(k)**: hydraulic conductivity (flow velocity)
- Drainage porosity **(μ)**: pore volume

Table 2: Some results for total porosity and permeability

Porous rocks	Total porosity (%)	Permeability (m/Day)
Sand and gravel	25 à 40	10 à1000

Fine sand	30 à 35	0,1 à 100
Clay	40 à 50	< 0,1
Chalk	10 à 40	1 à 100
Limestone (fissured)	1 à 10	< 1

I.4.1. Experimental determination of permeability

Experimental determination can be carried out in the laboratory or in situ:

- **In the laboratory:** the determination can be made using an undisturbed (unmodified) soil sample and applying a constant load. The permiability is then given by the Darcy (1856) formula:

$$\frac{Q}{A} = -K\frac{\Delta h}{L}$$

With :

- Q: filtering volume flow (m³/s).
- K: the hydraulic conductivity or "permeability coefficient" of the porous medium (m/s), which depends on both the properties of the porous medium and the viscosity of the fluid.
- A: surface area of the section studied (m²)
- $\Delta H/L$: The hydraulic gradient ($i = \Delta H/L$), where ΔH is the difference in piezometric heights upstream and downstream of the sample, L is the length of the sample.

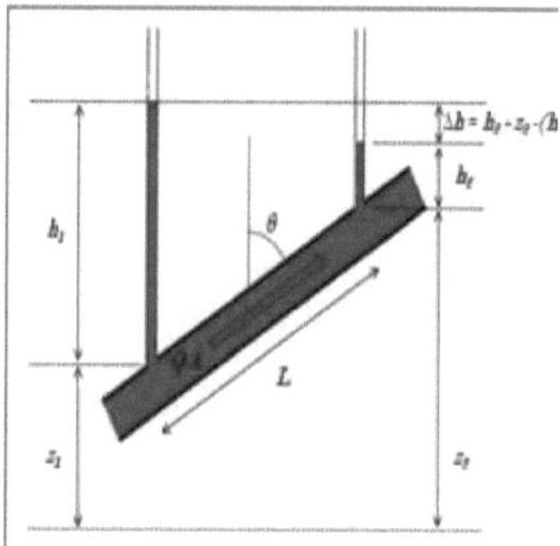

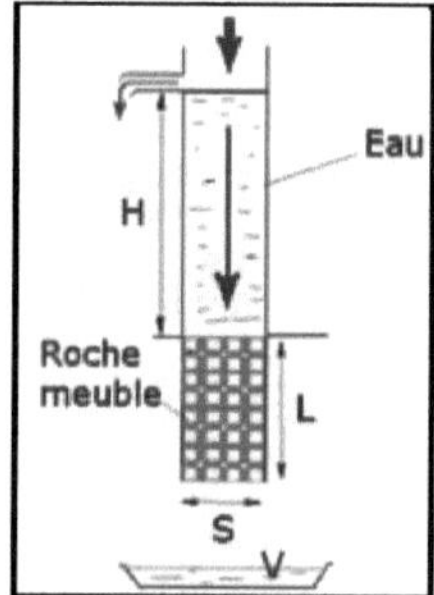

- **Insitu:** permeability can be determined by several methods:
  - a. **Well and piezometer method (saturated soil):** consists in digging a well in the water table at ϕ = 20 cm and depth P = 2m, in which

continuous pumping is applied and the fluctuation in water table levels is measured using one or more piezometers.

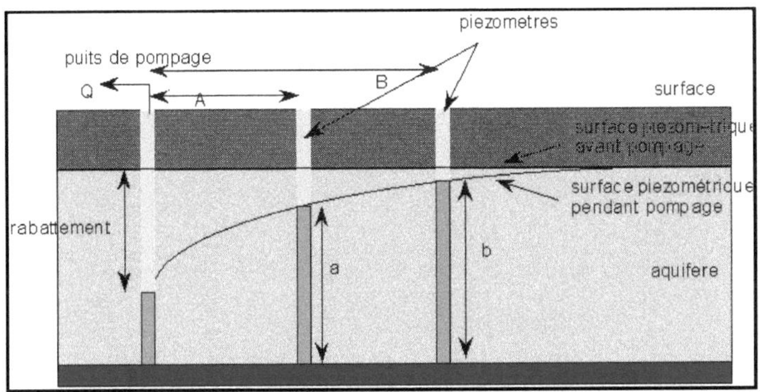

Diagram 3: location of a piezometer in a free-surface aquifer

Base = bedrock
- *If P(Assise Imper) ≤ 2m → a single piezometer is used*
- *If P (A.I) > 2m, two or more piezometers are used.*
- Always the depth of the well = trap depth = 2m

b. PORCHER's method

The Porchet soil water infiltration test consists of :

- Dig a hole of specified diameter, either with a hand auger (preferably 150 mm diameter), or with a spade (square hole 30 cm by 30 cm; width *l*), to a depth of 70 cm. This 70 cm depth is considered to be the infiltration depth in the case of soil infiltration of wastewater (filter trenches or spreading beds).

- For a period of 4 hours, use a hose or water bottles to maintain the water level at 25 cm above the bottom of the hole, i.e. 45 cm above the surface (height *h* below). The purpose of this operation is to return the soil to the water-saturated conditions that would be observed when operating a sewage treatment plant.

After 4 hours, measure (using a graduated water bottle, for example) the quantity of water to be added to keep the water level constant ($h = 25$ cm from the bottom of the hole or 45 cm from the surface) for 10 minutes.

c. Müntz method

The Müntz infiltrometer method is based on the principle of constant-load infiltration. A graduated reservoir maintains a constant water level of 30mm in a cylinder implanted in the ground. Time-dependent variations in the water level in the graduated feed tank determine the infiltration rate.

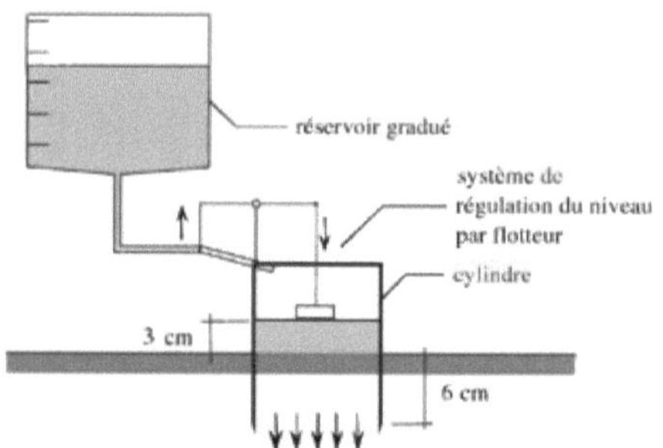

Figure 4: Müntz infitrometer

d. *DOUBLE RING trial*

This simple test is particularly well-suited to assessing surface and shallow-depth permeability (in the case of pavements with a reservoir structure), given its principle and measurement range. It consists in placing 2 concentric cylindrical rings (Ø 20 to 80 cm) on the ground, and sealing them with the soil by sinking them in 5 to 10 cm.

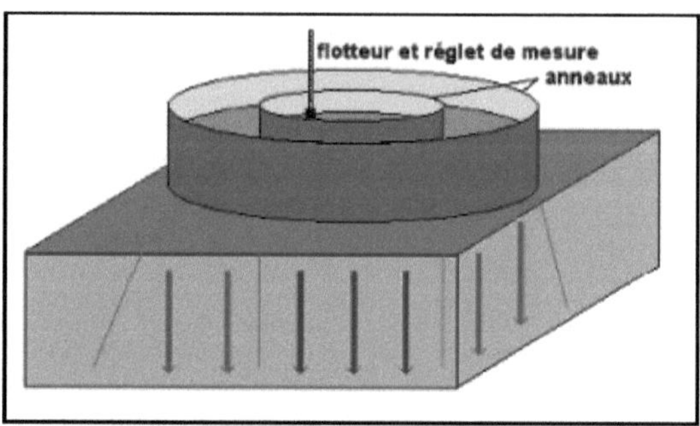

Figure 5: Diagram of test

(Source: CETE Nord - Picardie, guide chaussée réservoir 2002)

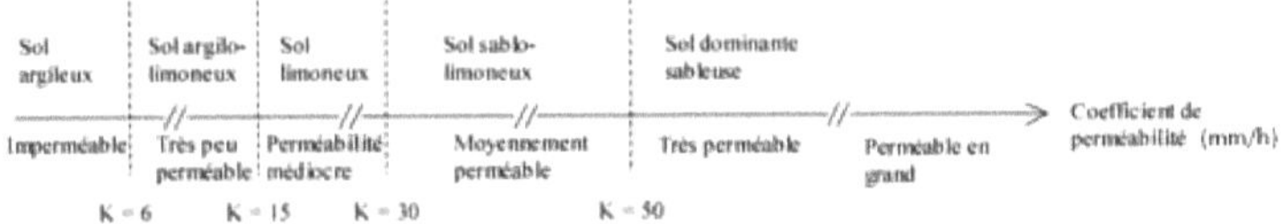

Figure 6: evolution of the permeability coefficient

I.4.2. Determination of drainage porosity

Drainage porosity (μ) is the ratio of the total volume of water particles flowing by gravity through the soil sample in question, to the total volume of the same sample.

The amount of water remaining in the sample can also be used to estimate a soil's **retention capacity**.

The **kinematic porosity** of a porous medium corresponds to the ratio between the volume of water flowing freely through the pores of the sample in question, and the total volume of the sample.

In practice, it is assumed that **kinematic** and **drainage porosity are similar**.

They are therefore grouped together under the term **"effective porosity"**.

Drainage porosity is therefore the ratio between the volume of water extracted from the sample by gravity (Ve) and the total volume of the sample (Vt).

$$\mu = \frac{Ve}{Vt}$$

Soil permeability and drainage porosity vary in the same direction.

Chapter II: Agricultural drainage systems and materials

II.1 Introduction

To remedy the phenomenon that causes soil losses due to prolonged saturation or excessive salinity, it is necessary to **consider systems for evacuating harmful water** even before it infiltrates the soil in the case of runoff, or after infiltration or an existing water table rises in the case of groundwater.

There are several drainage systems that have evolved over time. The most commonly used drainage systems in Tunisia are :

- **Surface drainage** to convey runoff (surface water) upstream and to the right of the areas to be treated. It also aims to eliminate any accumulation of water on the surface, as well as hypodermic runoff, within a reasonable timeframe for plants (less than 24 hours):
  - *Creation of colature ditches*
  - **Flow recalibration**
  - **Irrigation system**
- **Underground drainage:** underground drainage is a drainage technique designed to evacuate gravity-fed water and lower the water table to an optimum level for plant growth, towards a natural outlet (sea, wadi, sebkhat etc...). In practice, underground drainage should be carried out when the water table is less than 0.7 meters above the soil surface during the year
  - *Open ditch drainage*
  - *Underground pipe drainage*
  - *Underfloor drainage*
  - *Associated systems*   *Pipe + ditch*
    Management + tillage
    Pipe + subsoiling
  - *Well drainage*

In many cases, it is desirable to wrap drainage pipes in filter material (*filters and wraps*). These materials offer the following advantages:

- they prevent the penetration of soil particles (protection against *clogging*)

- they increase the circumference of the pipe

- they ensure better permeability all around the pipe

- they protect the pipe during transport and installation.

Filters are made up of gravel of various sizes in the 4 to 40 mm range, which can be classified into 3 sub-classes: (4/15, 15/25 and 25/40).

The use of this different class depends on the nature of the soil where the drains are to be used:

- 4/15 is used for heavy soil (clay).
- 15/25 is used for loamy soil.
- 25/40 is used for sandy soil.

This gravel filter facilitates the flow of water around the drain by increasing the contact surface.

There are two types of coating: thin and thick. Thin coating is less than 1cm thick. This normally consists of surrounding the pipe with a material of polyester origin at the manufacturing plant. Thick coatings are of the same type as thin coatings, but are wrapped around the drain at the laying stage

II.2 Surface drainage systems

II.2.1. Settling ditches

Ditching systems will be used ***prevent hillside runoff from spreading onto the plains*** to cause saturation or stagnation.

The purpose of these ditches ***is to divert runoff water*** from topographically higher areas.

These ditches may be on the outer limits of the area to be drained, or they may cross this area.

Figure 7: Colature ditches

These ditches will have a *trapezoidal cross-section* (for stability reasons) for which the slope of the embankment depends on the nature of the soil (sandy or heavy = clay).

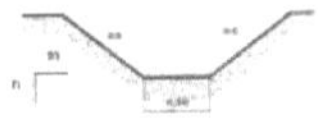

- Clay soil: m=2 and n = 3
- Clay-loam soil: m= 3 and n= 2
- Sandy soil: it is not advisable to have a ditch without protection.

This type of drainage is found in the Kalaat El Andalous, Outik and Sejnene plains

II.2.2. Flow recalibration

The principle of *recalibration is to increase the flow rate in the minor bed by increasing the flow cross-section through bed widening, deepening or both*

The recalibration of runoff concerns runoff that crosses the plains and causes stagnation and overflow.

II.2.3. Irrigation system

The irrigation system contributes to soil drainage when it avoids saturation of the root zone, and this can be ensured by respecting the irrigation time and doses corresponding to each crop.

II.3. Underground drainage systems

II.3.1. Open ditch drainage

Open ditches come in a variety of shapes, from triangular to trapezoidal. Various factors influence the hydraulic functioning of a ditch: (longitudinal slope, hydraulic radius, permeability of the soil in which it is dug, nature of the walls, embankment, etc.).

The role of ditches is to collect both surface water and groundwater and convey them to an outlet, which may be a larger ditch or an outfall.

Ditches should have a minimum depth of 80 cm, and their width should be calculated so that, during periods of heavy rainfall, the water level remains below 60 cm of the reference level. This method is generally chosen for draining forest, woodland or peatland soils.

Figure 8: Open ditch

The **advantages** of the open ditch :

- Easy to maintain (because it can be seen with the naked eye).

The disadvantages of this drainage system are :

- Soil loss
- Difficult access for working the soil, especially on clay soils.
- The appearance of halophilic plants (reeds) which slow down the flow.
- Requires a track for maintenance and cleaning.

II.3.2. Underground pipe drainage

Like ditches, buried pipes do not have the capacity to collect surface water. Their action is mainly directed at groundwater.

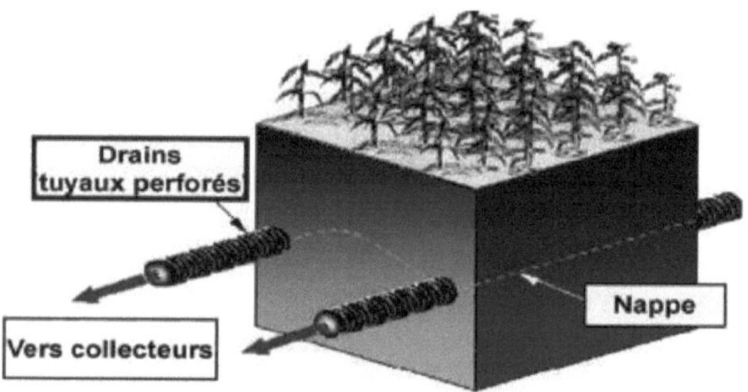

Figure 9: buried pipes

Dewatering is carried out by criss-crossing the plot with a primary network of drains buried at 1.5 m and spaced between 5 and 100 m apart, depending on the topographical configuration and nature of the terrain, protected by a layer of gravel (gravel filter ϕ 4/15, ϕ 15/25 and ϕ 25/40 mixed with sand to increase permeability) and surrounded by non-woven geotextile to prevent clogging and facilitate flow around the drains in the case of clay soils. The flow to be evacuated is connected to a main drain located along the most downstream side of the plot. The thickness of the membrane covered by the geotextile depends on the granulometry of the soil (PVC).

Figure 10: Coated agricultural drain

Figure 11: Perforated agricultural drain 200mm and 45m of geotextile

This system has gone through several types of pipe, starting with pottery, smooth PVC and finally corrugated PVC etc...

a. **Pottery pipes**

Pottery drains, which were almost systematically used before the advent of plastic drains, have the disadvantage, in addition to heavier handling and installation requirements, of offering only a low opening density (water enters the drain at the junction of the elements). Concentration favors the appearance of higher hydraulic gradients, so that the risk of soil particles being carried along and clogging the drain can become significant.

 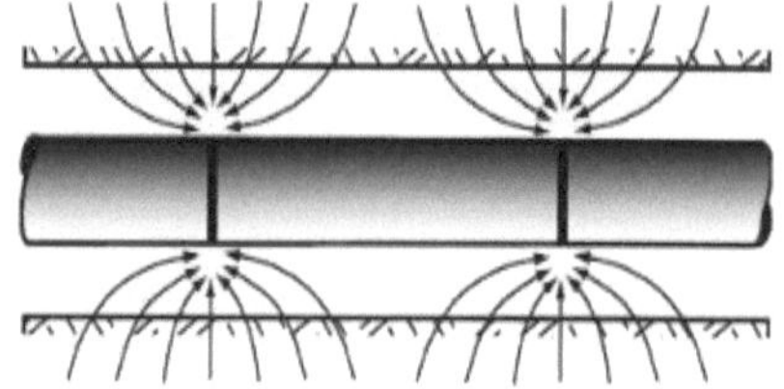

Figure 12: Pot pipes and their junctions

In Tunisia, we started using this type in 1960. Water penetrates all interstices. The most commonly used diameter is ϕ **50**. These pipes are formed by **30 cm** elements and laid at depths **≤2 m** by a machine called a trencher drainer with a clearance of **2** mm

Water entry surface (pot): Se = 3x∏xDxe = 3x∏x50x2 = 942 mm²/ml **≈ 10 cm²/ml**

The advantages of this system :

- Easy to manufacture.
- Resistance to crushing forces during installation and operation.
- Cheaper than PVC drains.
- Lifespan> 50 years in soil.

The disadvantages of this drainage system are :

- Heavy to transport.
- Fragile during installation and transport.
- Even more time-consuming to install.
- When the ground is unstable, there may be an offset between the elements.

b. Smooth PVC pipes

Water can enter this type of pipe via two types of cast iron: longitudinal (depending on length) or transverse (depending on radius). The dimensions of these cast irons are 3 cm long, 1 to 2 mm thick and spaced between 3 and 4 cm apart.

This type of drainage has a number of drawbacks, including deformation of the slots under the forces applied by the embankment, leading to deformation of the cross-sections and changes in flow rates that do not respect the slope studied. This type of drainage is not used in Tunisia.

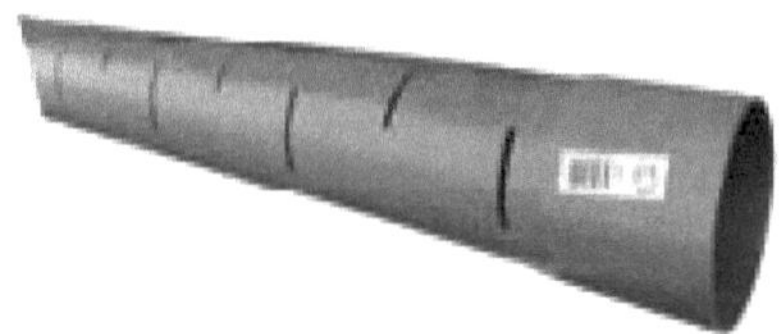

Figure 13: smooth PVC pipes

c. PVC corrugated pipes

To overcome the disadvantages of smooth PVC, drainage projectors have developed ringed PVC drains with 6 rows of slots. The slots for this type of drain are elliptical in shape (e=2mm and d=4mm) implanted in the inner rings and spaced 2 cm apart. These pipes are coated with a geotextile filter to prevent clogging. The pipes are sold in 100m coils. They are easy to lay and transport, avoiding the problem of soil compaction during laying or transport.

-   ***Water entry surface (PVC)***: Se = Number of gnawed x Number of slots x d x e

    $$Se = 6 \times 50 \times 4 \times 2 = 2400 \text{ mm2} \approx \mathbf{24 \ cm^2/ml}$$

Figure 13: PVC corrugated pipes

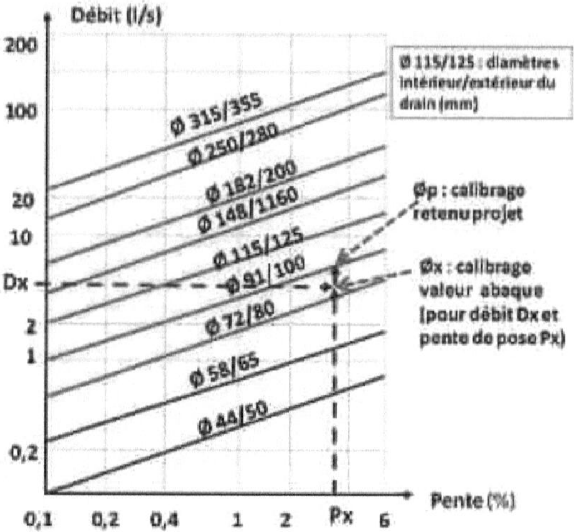

Figure 14: Flow chart for PVC corrugated drains (from manufacturers' flow charts: Oltmanns & Wavin)

II.3.3. Drainage by subsoiling

Drainage by subsoiling (furrowing) involves digging slots in the soil in the form of *furrows spaced 1 m apart, 0.7 m deep and 5 cm thick, using a towed subsoiler.* This network of furrows must be oblique or perpendicular to the buried pipes to facilitate flow to the drains. This type of drainage is generally used for heavy soils, and is combined with buried pipe drainage to facilitate the flow of surface water towards the pipes. This type of drainage is suitable for clay soils and must be redone every 3 years.

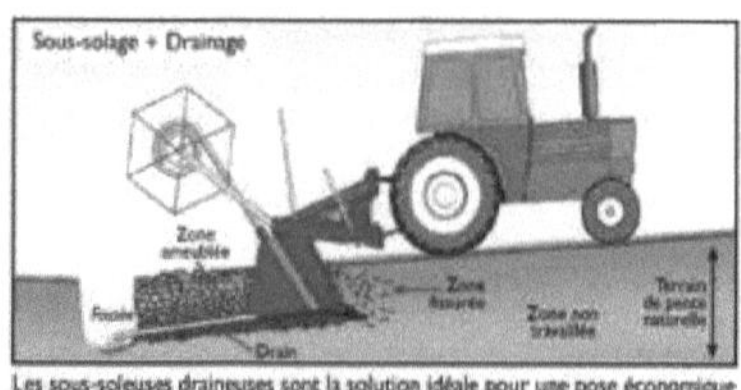

Les sous-soleuses draineuses sont la solution idéale pour une pose économique des drains de Ø 50 à 125 mm, dans les terrains de pente naturelle. Pose de tuyaux souples, câbles électriques, câbles téléphoniques...

Figure 15: Subsoiler

II.3.4. Drainage by associated systems

This type of drainage can be a combination of two or more systems, depending on the nature and occupation of the soil:
- Pipe + ditch
- Colature + ditch
- Tillage + pipe + ditch

II.3.5. Well drainage

Well drainage consists of digging a series of wells to lower the water table so as to create a root zone outside the water table (1.6 m for annual crops and 2 m for shrubs). The diameter of these wells is >1 m.

II.4. Advantages and disadvantages of agricultural drainage

II.4.1. Drainage advantages

The advantages of field-scale drainage can be summarized as follows

- It promotes the beneficial activity of soil bacteria and improves soil condition.
- There is less surface runoff and soil erosion on drained land.
- Improved trafficability reduces structural damage to the soil. Soil compaction is reduced and less energy is required for on-site operations. Drainage also enables faster field operations. As a result, the growing season can be extended and crops can reach full maturity.
- Crop yields increase due to improved water management and plant nutrient uptake.
- Higher-value crops can be planted, and new and improved cropping systems can be introduced.
- Drainage maintains favorable saline and atmospheric environments in the crop root zone.

IV.4.2. Disadvantages of agricultural drainage

- Hydrological effects: The only sometimes negative hydrological effects of drainage are attributed by some authors to the oversizing of drainage ditches and sewerage systems. The evacuation of drainage water often leads to the deepening of ditches. As a result, floods are evacuated more quickly, with potentially negative effects downstream.

- Maintenance: One of the most inconvenient things about drainage systems is maintenance. It's a must, because if the drainage system is clogged, it won't work 6 properly. You need to check for blockages and debris that could block the flow of water.

- Water quality degradation: drainage water contaminated with salts and trace elements is generally discharged untreated into a natural outlet.

Chapter III: Agricultural sanitation principles and methods

III.1. Introduction

If remediation is the action of cleaning up, farmers do it to allow soils to reach their maximum production capacity. Let's take a look at the principles and methods of sanitation from this perspective.

III.2 Production and sanitation

Farmers want to use the growing season to its full potential in terms of solar radiation and heat to produce plants associated with crops. They want to maximize or optimize their income by seeking maximum or optimum yields. In some cases, they are looking for income stability or production stability because they have contracts to fulfill or need plant products to feed their animals. In all cases, they want to minimize the vagaries of the weather due to excessive rainfall, insufficient rainfall or poorly distributed rainfall. They also want to be able to carry out the necessary work, such as sowing and harvesting. There's no point in producing if you can't harvest. Farmers carry out water purification and other work in line with the objectives described.

III.3 Sanitation objectives

There are three main objectives to soil remediation from a hydraulic point of view:

- Remove runoff water to allow access to the field and reduce plant flooding.
- Lower the water table to allow plant roots to grow under favorable conditions, and to allow machine traffic for operations such as tilling, sowing, spraying, weeding, fertilizing and harvesting. Operations must be carried out in favorable conditions and without damaging the machinery or the soil.
- Supply water to plants during periods of water deficit, so that they don't suffer from drought.

III .4 Techniques

The following hydraulic sanitation techniques are classified according to sanitation objectives:

1- **Remove run-off water:**
   - Shallow ditches (30 cm);
   - Deep ditches (90+ cm);
   - Gutters
   - Round boards and stripes

- Watercourses.

2- Lower the tablecloth:

- Underground drainage ;
- Deep ditches (90+ cm);
- Mole drainage.

3- Getting water to the plants:

- Controlling the water table ;
- Underground irrigation
- Sprinkler irrigation ;
- Surface irrigation
- Drip irrigation.

The first four means of removing runoff are at field level and the responsibility of the farmer, while the last is at regional or watershed level and the responsibility of the local authority. Some techniques, such as deep ditches, meet more than one objective. The installation of drains is used for underground drainage, water table control and underground irrigation.

III.5 Approaches

When analyzing and solving sanitation problems, it is necessary to follow a procedure. This involves the following steps:

1. Identify the moisture problem by observing the following symptoms:

   - Presence of a water table ;
   - Condition of plants showing water stress problems ;
   - Plant types characteristic of wetlands (cattails, etc.);
   - Presence of a slick, periods of its presence, duration of its presence;
   - The circumstances surrounding the occurrence of the problem.

2. Sources of the problem :

   - Soil type (compact, poorly structured)
   - Pedology and geology (soil depth, type of formations) ;
   - Presence of an indurated layer
   - Low hydraulic conductivity
   - Geomorphology of the terrain (basin, depression)
   - Flat land with no natural outlet.

3.   Identifying solutions. Solutions must be identified in relation to the problem and its sources, and in relation to economic conditions.

It's never a good idea to prejudge the solution before studying the problem.

Chapter IV: Sizing an agricultural drainage network

IV.1 Introduction

The sizing of a drainage network consists in determining the following parameters: drain spacing, unit flow rate, characteristic flow rate, maximum flow rate and maximum length. It should be noted that the formulas used for sizing a drainage network are largely empirical, based on observation, experience and statistics. For this reason, when making calculations, the operator must homogenize the units he is working with to avoid encountering aberrant results.

VI.2 Determining drain spacing

The distance separating the 2 axes of the drain is called the drain spacing (E). The determination of this distance is based on experience adjusted by theory. There are 2 methods: the steady-state method and the variable or transient method.

A. Steady-state method

This regime consists of **maintaining** the level of a water table just below the root zone at a constant level.

This method is based on the assumption that *the inflow to the root zone is equal to the outflow from the drains*, to guarantee a constant water table level below the root zone.

Drain spacing depends on soil permeability (K), depth of impermeable bedrock, rain intensity I_0 and crop rooting depth.

The spacing will be calculated for both *buried pipe* and *open ditch* drainage systems.

    a. Drains resting on impermeable base P < 2m (P=P$_d$) :

    i. Trench backfill more permeable than the soil in place (K<10^{-6} m/s: clay soil)

:

The trench backfill is said to be more permeable than the soil in place if soil other than that in place is used as backfill, or if the drains are in ditches.

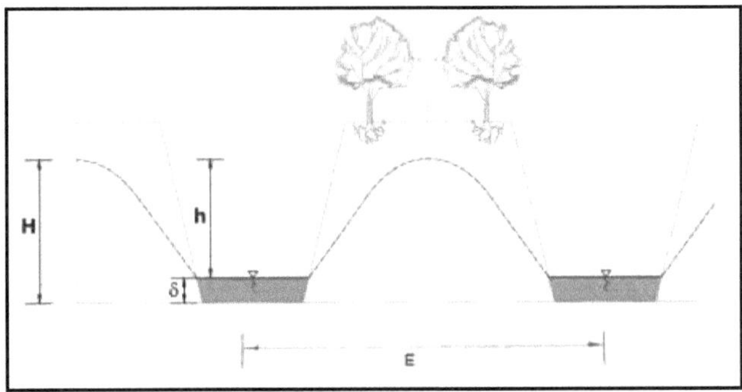

Figure 16: Drains resting on impermeable bedrock

The basic formula for the steady state is the GUYON formula.

- **Homogeneous, isotropic soil:** has the same physical characteristics in all directions.

$$\frac{IE^2}{4} = K.H^2\left(1 - 2R\frac{I}{K}\right) - \delta^2 K \quad \text{(1)}$$

I: Filtration rate in (mm) or (m^3/ha).

K: permeability of drained soil in (m/s)

E: Distance between drains (m)

H: Depth difference between the surface of the lowered water table and the impermeable bedrock (in our case, the depth at which the drains were laid, as P < 2m).

R: Dimensionless coefficient (for calculation R = 0.25).

δ Height of water in the drain. For homogeneous, isotropic soil, the filtration rate (I) is calculated as follows:

$$I = \sigma.I_0$$

I: Filtration rate in (mm) or (l/s/ha): *portion of rainfall that infiltrates*.

σ Correction coefficient (restitution) dependent on permeability K and slope i.

I$_0$: Drained flow in (mm) or (l/s/ha). *The amount of rain that falls*. This is also the critical rain intensity that causes stagnation in the root zone (3 days).

The values of σ are given in the following table:

31

K (m/s) \ pente (^{0}/oo)		1^0/oo		5^0/oo		10^0/oo		15^0/oo
K> 5 .10^{-6} (Forte)	1	1	0.9	0.8	0.8	0.7	0.7	0.6
10^{-6}< K <5 .10^{-6} (moyenne)	1	1	0.9	0.8	0.7	0.6	0.6	0.5
K< 10^{-6} (faible)	0.9	0.9	0.8	0.7	0.7	0.6	0.6	0.5

If irrigation is used, the filtration flow rate is equal to the critical flow rate:

$$I = q_C$$

NB: In the case of drainage by buried pipes, the terms δ (the height of water in the drain) and **R** (the dimensionless coefficient) can be neglected because the value of the filtration rate **I** is too negligible compared with the permeability **K**.

This gives us the following formula:

$$\frac{IE^2}{4} = K \times h^2 \quad (2)$$

- **Homogeneous, anisotropic soil:**

$$\frac{IE^2}{4} = H^2 \times K_h \times \left(1 - 2R \times \frac{I}{K_v} \right) - \delta^2 \times K_h^{\,2} \quad (3)$$

Kh: Horizontal component of permeability.

Kv: Vertical component of permeability.

- **Heterogeneous soil :**

$$\frac{IE^2}{4} = H^2 \widetilde{K}_h(H) \times \left(1 - 2R \times \frac{I}{K_v(H)} \right) - \delta^2 \widetilde{K}_h(\delta) \quad (4)$$

ii. Trench backfill with the same permeability as the soil in place (K>10$^{(-6)}$) m/s) medium to high permeability

- **Homogeneous and isotropic soil :**

$$\frac{IE^2}{4} = \gamma \times K \times h^2 \qquad (5)$$

- **Homogeneous, anisotropic soil**

$$\frac{IE^2}{4} = \gamma \times K_h \times h^2 \qquad (6)$$

- **Heterogeneous soil :**

$$\frac{IE^2}{4} = \gamma \times \widetilde{K}_h \times h_m \qquad (7)$$

hm: Average height in different soil layers
γ Coefficient dependent on h and E.

$$\Rightarrow Si \ \frac{h}{E} \leq 0.1 \Rightarrow \gamma = 1$$

$$\Rightarrow Si \ \frac{h}{E} > 0.1 \Rightarrow \gamma = 1.2 - \frac{2h}{E}$$

b. Drains do not rest on impermeable base p > 2m :

i. Trench backfill more permeable than existing soil :

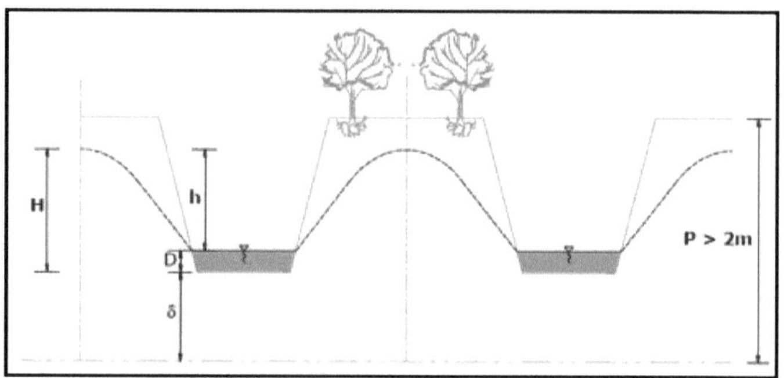

Figure 17: Drains not resting on impermeable bedrock

$$H = D + h$$

ERNEST formula:

- **Homogeneous and isotropic soil :**

$$\frac{IE^2}{4} = \frac{K.h^2}{\dfrac{1}{1 + 2 \times \left(\dfrac{\delta + D}{h}\right)} + \dfrac{4h}{\pi \times E} Ln\left(\dfrac{\delta + D}{\phi}\right)} \qquad (8)$$

ϕ Wetted perimeter of drain = inside diameter of drain

NB: to simplify calculations, values for filtration rate I and permeability K should be translated into the same unit (e.g. m/d).

- **Homogeneous, anisotropic soil:**

$$\frac{IE^2}{4} = \frac{K_1.h^2}{\dfrac{1}{1 + 2 \times \dfrac{K_1}{K_2}\left(\dfrac{\delta + D}{h}\right)} + \dfrac{4h}{\pi \times E} \times \dfrac{K_1}{K_2} Ln\left(\dfrac{\delta + D}{\phi}\right)} \qquad (9)$$

ii. Trench backfill has the same permeability as the soil in place :

To carry out these calculations, you must first : Converting the real schematic into a fictitious one that facilitates calculations:

δ': Equivalent height, which must be less than δ

δ' : Equivalent water height, determined by the following formulae :

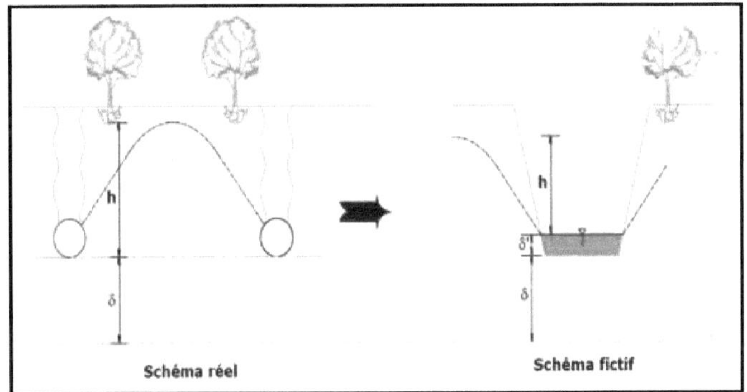

Figure 17: transformation of real schematic into fictitious schematic

GUYON formula:

$$\frac{\delta'}{E} = \frac{\dfrac{\delta}{E}}{1 + \dfrac{4 \times \delta}{E}\left(1{,}1 + \dfrac{20 \times h}{E}\right)} \qquad \text{(A)}$$

NB: This formula (A) is associated with another formula which gives the gap (E):

$$\frac{IE^2}{4} = \gamma \times K \times h^2 + 2 \times K \times h \times \delta' \qquad \text{(10)}$$

Iterative calculation to determine δ' so that **Ef = Ec+/-0.1**

E fixed	δ(formula A)	Ec value

HOOGHOUDT formula

$$\frac{\delta'}{E} = \frac{\dfrac{\delta}{E}}{\left(1 - \dfrac{\delta}{E}\sqrt{2}\right)^2 + \dfrac{8 \times \delta}{\pi \times E}\left(\dfrac{\delta\sqrt{2}}{\phi}\right)} \qquad \text{(B)}$$

With this formula we have the following formula which gives the gap E :

$$\frac{IE^2}{4} = 2 \times K \times h \times \delta' \qquad (11)$$

B. Variable dry-run method :

This system consists of lowering the initial level of a "troublesome" water table to a final level, within a period not exceeding the tolerance time of the crops planted in the drained soil.

a. Drains resting on impermeable base p < 2m :

i. Trench backfill more permeable than existing soil :

The basic formula for the variable drying regime is the **GUYON** formula:

- **Homogeneous and isotropic soil :**

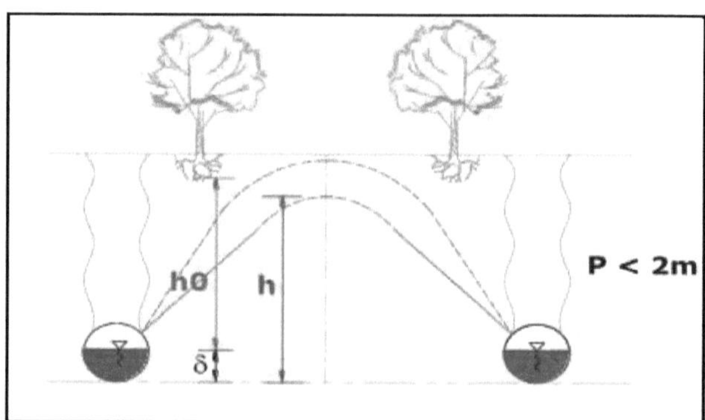

**Figure 18: Drains resting on impermeable bedrock
(Trench backfill more permeable than existing soil)**

$$E^2 = \frac{4\delta}{N \times Ln\left(\frac{1+2\frac{\delta}{h}}{1+2\frac{\delta}{h_0}}\right)} \times \left(\frac{K \times t}{\mu} - 2R(h - h_0)\right) - \frac{4\delta^2 K}{N} \qquad (12)$$

With :

N: dimensionless coefficient = 0.435

R: Dimensionless coefficient (for calculation R = 0.25).

μ : drainage porosity (given by abacus n°1) or by *the following formula* :

$$\mu = 4,7008 \times Ln\,(K) + 6,2016 \quad avec\ K\ en\ (cm/h)$$

t: drawdown time: crop tolerance time.

If K(m/s) then t(s)

If K(m/j) then t(j)

- **Homogeneous, anisotropic soil:**
  - *Open ditches*

$$E^2 = \dfrac{4\delta}{N \times Ln\left(\dfrac{1+2\,\delta/h}{1+2\,\delta/h_0}\right)}\left(\dfrac{K_h \times t}{\mu} - 2R\dfrac{K_h}{K_v}(h-h_0)\right) - \dfrac{K_h}{K_v}\dfrac{\delta^2}{N} \qquad (13)$$

  - *Buried pipes*

For buried pipes => **(12)**, a restriction is allowed which can be written as :

$$E^2 = \dfrac{2 \times K \times h_0 \times h \times t}{\mu \times N \times (h_0 - h)} \qquad (14)$$

- **homogeneous, anisotropic soil:** for this type of soil, the R and δ terms are neglected:

$$E^2 = \dfrac{2 \times K_h \times h_0 \times h \times t}{\mu \times N \times (h_0 - h)} \qquad (15)$$

  - *Buried pipes*

$$E^2 = \dfrac{2 \times \widetilde{K}_h(h_m) \times h_0 \times h \times t}{\mu \times N \times (h_0 - h)} \qquad (16)$$

Avec :

$$h_m = \dfrac{h + h_0}{2} \qquad\qquad V_m = \dfrac{h_0 - h}{t} \qquad\qquad w = \dfrac{\delta}{h_m}$$

ii. Trench backfill has the same permeability as the soil in place :

- **Homogeneous and isotropic soil :**

$$E^2 = \frac{2 \times K \times h_0 \times h \times t}{\mu \times N \times (h_0 - h)} \qquad (14)$$

- **Homogeneous, anisotropic soil:**

$$E^2 = \frac{2 \times K_h \times h_0 \times h \times t}{\mu \times N \times (h_0 - h)} \qquad (15)$$

b. Drains not resting on impermeable base p > 2 m :

i. Trench backfill more permeable than the soil in place: No formulas

ii. Trench backfill more permeable than existing soil :

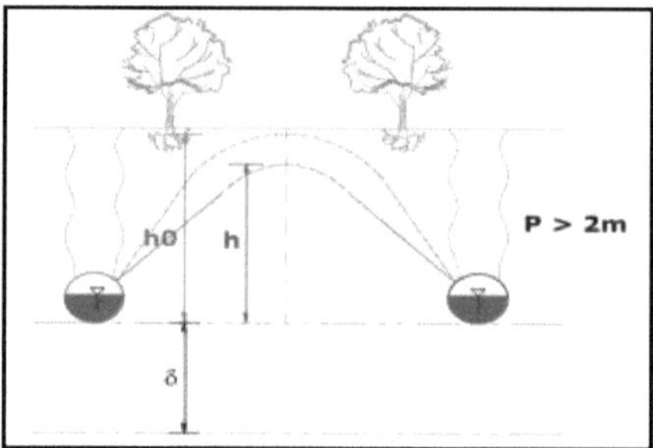

Figure 19: Drains not resting on impermeable bedrock

$$E^2 = \frac{4 \times K \times t \times \delta'}{\mu \times N \times Ln\left(\dfrac{1 + 2\delta'/h}{1 + 2\delta'/h_0}\right)} \qquad (17)$$

The value of δ ' is given either by formula **(A)** or by formula **(B)**

(A) and **(17)** or **(B)** and **(17)** ==> We can also use the abacuses n°2: We take a reading of the value of ρ and we directly determine the gap **E** by the formula :

$$E = \frac{h_m}{\rho}$$

hm: Average height of water in different soil layers (m) :

$$h_m = (h + h_0)/2$$

IV.3 Determining drain flow rates

IV.3.1. Determination of characteristic flows and unit flows

There are two types of flow rates:

Characteristic flow rate q_c: flow rate per unit area drained (l/s/ha).

Unit flow rate q: flow rate per unit length of drain (m³/s/ml).

$$\begin{aligned} Q &= q \times L \\ Q &= q_c \times E \times L \end{aligned} \quad \Rightarrow \quad q_c = \frac{q}{E}$$

A. Steady-state method :

I0: Portion of rain that falls (mm/d).

I: Portion of rainfall that infiltrates (mm/d).

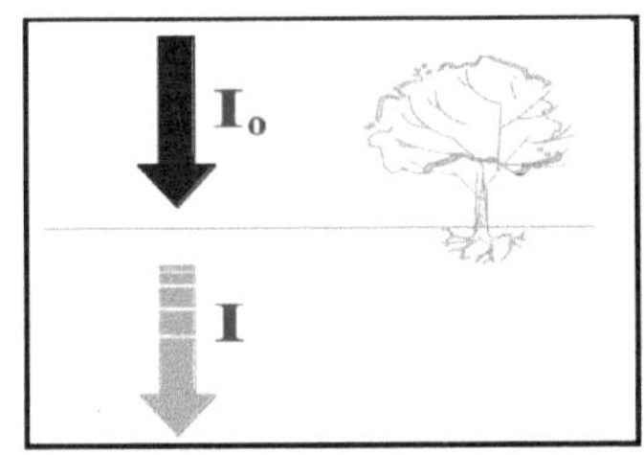

$$I\,(mm/j) = (m^3/j/ha) => q_C$$

$I = q_c$: for all steady-state cases.
$$q_c \text{ (mm/h)} = I \text{ (mm/h)}$$
$$q_c \text{ (l/s/ha)} = 116 \times I \text{ (m/d)}$$
I: can also be the irrigation intensity.

So for steady state :

$$q = I \times E$$

q (m³/s/ml), E(m) and I (m/s)

The distance **E** is determined by the formulas in **(1)** to **(11)**.

B. Variable dry-run method :

Determine **q** and derive q_c

1st case :

If it plut $q_c = I_0$ / **asphyxiation plant tolerance time**.

2th case :

$$q_c = \frac{q}{E}$$

And **q** varies from <u>case to case</u>:

a. Drains resting on impermeable base p < 2m :

i. Trench backfill more permeable than existing soil :

- **Homogeneous and isotropic soil :**

$$q = \frac{2 \times P \times h_m \times K \times (h_m + 2.\delta) \times E}{N \times E^2 + 4R(h_m + \delta)^2} \qquad \text{(q13)}$$

The gap E is determined by formula **(13)**

P: dimensionless coefficient = 0.73

N: dimensionless coefficient = 0.43.

hm: Average height in different soil layers (m)

K: permeability of drained soil in (m/s)

ii. Trench backfill has the same permeability as the soil in place :

- **Homogeneous and isotropic soil :**

$$q = \frac{2 \times P \times h_m^2 \times K}{N \times E} \qquad \text{(q14)}$$

With **E**: determined by formula **(14)**.

b. Drains not resting on impermeable base p > 2m :

i. Trench backfill more permeable than the soil in place: No formulas

ii. Trench backfill has the same permeability as the soil in place :

$$q = \frac{2 \times P \times K \times h_m \left(h_m + 2\delta' \right)}{E} \quad \text{(q17A ou17B)}$$

The distance E is determined by formulas **(17) and (A)** or **(17) and (B)**.

IV.4. Determination of maximum flow rates in drains

The maximum flow (m³/s) per drain is calculated as follows:

$$Q_{Max} = 21{,}82 \times D_{int}^{8/3} \times i\left(\% o \right)^{1/2}$$

i: the slope of the drain

D_{int}: internal diameter of drain (m)

IV.5. Determining maximum drain lengths

The maximum length (m) per drain is calculated as follows:

$$L_{Max} = \frac{21{,}82 \times D_{int}^{8/3} \times i\left(\% o \right)^{1/2}}{q(m^3 / s / ml)}$$

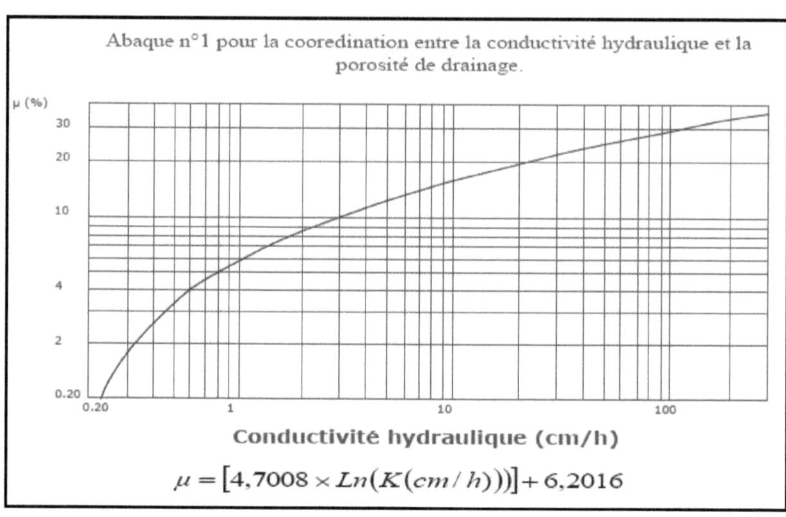

$$\mu = \left[4{,}7008 \times Ln(K(cm/h))\right] + 6{,}2016$$

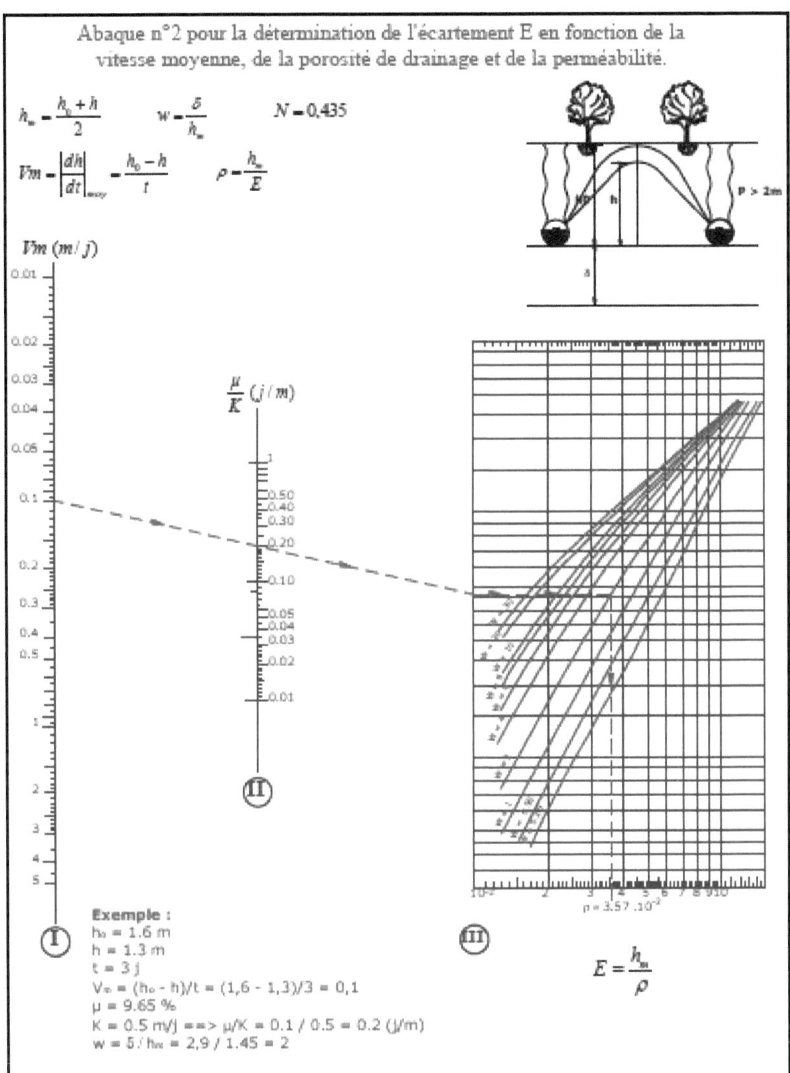

Abaque n°2 pour la détermination de l'écartement E en fonction de la vitesse moyenne, de la porosité de drainage et de la perméabilité.
$h_w = \dfrac{h_0 + h}{2}$
$w = \dfrac{\delta}{h_w}$
$N = 0,435$
$Vm = \left|\dfrac{dh}{dt}\right|_{moy} = \dfrac{h_0 - h}{t}$
$\rho = \dfrac{h_w}{E}$
$Vm\ (m/j)$
$\dfrac{\mu}{K}\ (j/m)$
Exemple :
$h_0 = 1.6$ m
$h = 1.3$ m
$t = 3$ j
$Vm = (h_0 - h)/t = (1,6 - 1,3)/3 = 0,1$
$\mu = 9.65\ \%$
$K = 0.5$ m/j ==> $\mu/K = 0.1 / 0.5 = 0.2$ (j/m)
$w = \delta / h_w = 2,9 / 1.45 = 2$
$\rho = 3.57 \cdot 10^{-2}$
$E = \dfrac{h_w}{\rho}$
P > 2m
I
II
III

Chapter V: Clogging drains and filter materials

V.1. Introduction

It's disastrous to find that after a drain has been installed, it becomes clogged and the investment is lost. Drain clogging is influenced by the characteristics of the soil, the drain and the installation conditions. To control drain clogging, it's important to understand the phenomena involved. Filter materials are used to control silting of drains, and it is necessary to be familiar with them and their conditions of use in order to make appropriate recommendations.

V.2 Types of clogging

Before describing the phenomena of drain clogging, it is important to understand the forms and origins in terms of terminology.

Drain pipe clogging can take two forms:

- **External clogging** is the total or partial obstruction of perforations and/or reduction in the hydraulic conductivity of the soil in the vicinity of the drain, which limits the penetration of water into the drain. The drain then loses much of its hydraulic efficiency.

- **Internal clogging** is the total or partial obstruction of the drain by soil particles, roots or chemical or biological deposits. This clogging leads to a reduction in the pipe's hydraulic cross-section and conveyance capacity.

Clogging may have a single cause or a combination of causes. The main causes are :

- **Mineral clogging**: this is caused by the migration of mineral particles that settle in the pipe (**internal clogging**) and/or become immobilized in the area around the drain. The latter leads to the formation of an area of low permeability (**external clogging**). This clogging may occur soon after laying, during the soil consolidation period in the trench near the drain. This is known as primary clogging. It is mainly due to poor installation conditions where the soil is very wet or saturated. Clogging can also occur during subsequent flow periods, when it is referred to as secondary clogging. Secondary clogging is mainly due to the nature of the soil. The secondary clogging of drains by soil particles is also known as drain silting. It is the main form of clogging that practitioners have to deal with. In some cases, drain silting can occur very quickly, even within a year of installation.

- **Physico-chemical" and "biological" clogging**: these are caused by modifications to the environment induced by the installation of the drains, leading to the proliferation of microflora adapted to the new conditions and/or deposits resulting from chemical transformations. Ferric" clogging is the most widespread type: it combines iron oxide deposits obstructing perforations and the development of a bacterial gel inside the drain.

- **Root clogging**: due to the accumulation of root hair in the drain. Root clogging occurs mainly in drain systems carrying water from a source. In this case, the drain is a privileged medium for root attraction, as it constitutes an easily usable reserve of water and air. Rootlets penetrate the drain through the perforations and, when they die, create plugs in the pipes that impede water flow.

V.3. Geotextile membranes

There are different categories of geotextile membranes.

V.3.1. Woven membranes

Woven membranes are made up of fibers oriented in two perpendicular directions and interwoven with one another. Compared with other manufacturing methods, weaving is a more expensive process, but it has the advantage of producing a product with a simple structure: pore size distribution is up to a point uniform, simple and easy to determine. What's more, the relatively simple geometry of woven membranes means that their mechanical properties can be directly related to those of the fibers.

It should be noted, however, that the stress characteristics of woven membranes are almost always presented in terms of the warp or weft direction, but if the membranes are subjected to stress in another direction (diagonal), their properties are considerably altered.

On the whole, woven membranes offer moderately high to very high resistances, and also have a simple pore structure.

V.3.2. Knitted membranes

Whereas in woven membranes the strands are essentially straight, knitted membranes are made up of loops of fibers connected by linear segments. Thanks to this structure, knitted membranes can be subjected to tension in one or more directions without significantly increasing the stress on the fibers. The knitting process has two advantages over weaving. It is less expensive, and it offers the possibility of manufacturing tubes. One application for these tubes is their use as filters around agricultural drains.

V.3.3. Non-woven membranes

This group includes all membranes that are neither woven nor knitted. They are made up of fibers bonded together by various processes, giving them special properties. On the whole, non-woven membranes are relatively inexpensive and have low to medium tensile strengths. They are also highly deformable. They are widely used as filters, drains, separating agents or in light reinforcement work.

V.3.4. Needled membranes

Needling is a mechanical process in which filaments are intertwined with needles, imparting a certain strength to the resulting mat. For greater strength, several layers can also be superimposed and needled together.

Needled membranes are thick relative to their weight (85-90% void), and the pore structure is quite complex. This can be an advantage in filtration.

V.3.5. Thermally bonded membranes

The fibers are bonded together by passing between two heated cylinders under high pressure. The filaments are welded together at the points of contact. The resulting membrane is relatively thin, and the configuration and size of the pores are independent of the stress applied to the membrane. However, it is often the case that if the fiber web is heated sufficiently to create a strong bond between the fibers, the result is a degradation of their mechanical properties and a reduction in their orientation.

V.3.6. Chemically bonded membranes

These membranes are produced by impregnating the fiber fleece with a resin that serves to bind them together. The thickness and structure of these membranes are intermediate between needle-punched and thermally bonded membranes. However, this method is the most expensive and, all things being equal, chemically bonded membranes have less void and lower permeability.

V.3.7. Other types

Membranes manufactured using a combination of these bonding techniques can also be found. For example, chemically bonded membranes are often needle-punched. On the other hand, many membranes are produced using more than one construction and bonding technique: for example. It is common practice to needle-punch fibers onto a woven backing. It is clear, therefore, that there is a wide variety of membranes, and it is also clear that an even wider range can be obtained with the development of new techniques and materials. The range of characteristics of these membranes is very wide, both in terms of pore characteristics and mechanical properties. Their service life can also be very different. The engineer will need to recognize these differences and select the most suitable membranes for each particular application.

Chapter VI: Identifying drainage problems

VI.1 Introduction

The engineer is often called upon to identify the causes of a malfunctioning underground drainage system. For the farmer, an underground drainage system presents a problem when the soil remains wet for varying periods of time. Beyond these appearances, every subsurface drainage problem presents symptoms that the engineer will need to observe to better identify the problem.

Underground drainage problems are all the more difficult to identify as they are buried with the drain and are not directly visible. Like a doctor, the engineer must try to detect all the symptoms that will enable him to identify the disease in the drainage system. Once the problem and its causes have been identified, he or she can recommend appropriate corrective measures, if possible.

The aim of this chapter is to present the various problems associated with malfunctioning underground drainage systems and the symptoms that may be associated with them. Knowing the problems and the symptoms associated with them, we will try to develop a strategy for identifying the problems.

VI.2 Operation of an underground drainage system

Before discussing how to identify problems in underground drainage, it's a good idea to describe how a normal system works and how it performs.

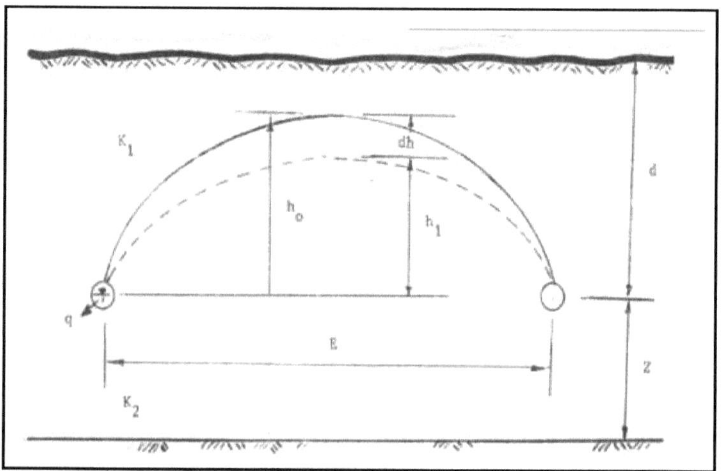

Figure 20: Diagram of an underground drainage system

An underground drainage system is characterized by (Figure 20)
* physical limits :

- drain depth "d

- the depth of permeable soil beneath the "Z" drains

- distance between drains "E

- drain radius "r

* soil properties :

- the hydraulic conductivities of the soil layers above and below the "K1 and K2"

drains

- equivalent drainage porosity

* hydraulic characteristics :

- the height of the water table above the "h0" and "h1" drains

- unit drain flow "q

The operation of an ideal underground drainage system assumes that the drains are installed in a trench whose hydraulic conductivity is greater than that of the surrounding soil (Ktrench > Ksoil).

For a normally functioning drainage system, the water table has a parabolic shape like the one shown in figure 13.2. The trench floor and drain offer very little resistance to water entry, and the water table almost reaches the drain. The hydraulic head near the drain is generally less than 20 cm.

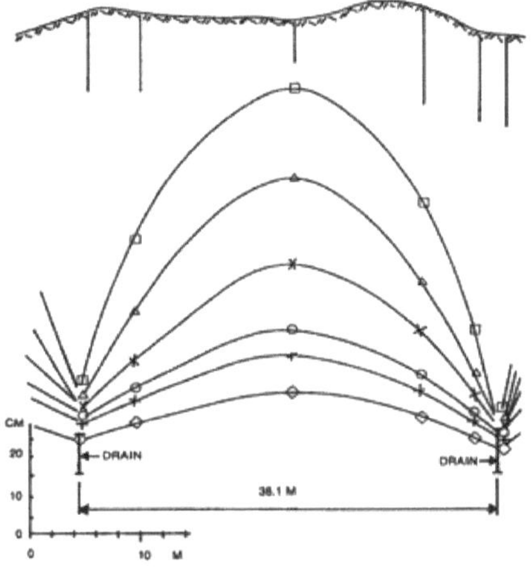

Figure 21: Profiles of the water table during drawdown

When the drain is not flowing under load, it is considered to be flowing on a free surface, and we can approximate the hydraulic slope to the slope of the drain; this is the case normally considered when designing. Note that it is not the slope of the drain that causes the flow in the drain, but the hydraulic slope of the water line above the drain

VI.3 Problems and symptoms

When a farmer mentions that his underground drainage system isn't working properly, it's because he believes his system isn't performing as expected. For him, the problem takes the form of delays in entering his field in spring, autumn or following heavy rain, traffic difficulties and, sometimes, plant growth problems and mediocre yields. The engineer's role is to distinguish drainage problems from other problems, to identify and solve them. Before defining an approach to the identification of different subsurface drainage problems, it would be useful to identify the problems likely to be encountered and to present the symptoms associated with them.

VI.3.1 Drain broken, crushed or blocked by a foreign body

A broken, crushed or clogged drain partially or completely reduces the drain cross-section. When the cross-section is completely blocked, water flows back upstream. Under the pressure created in the drain, the water diffuses into the ground, re-entering the drain at a point downstream of the break or obstruction (Figure 22). If the drain is working properly, the water table is drawn down into neighbouring drains. On flat ground, the situation upstream of the problem point is equivalent to the absence of a drain, and the drainage system behaves as if the spacing between the drains were double that installed.

a- Flat land

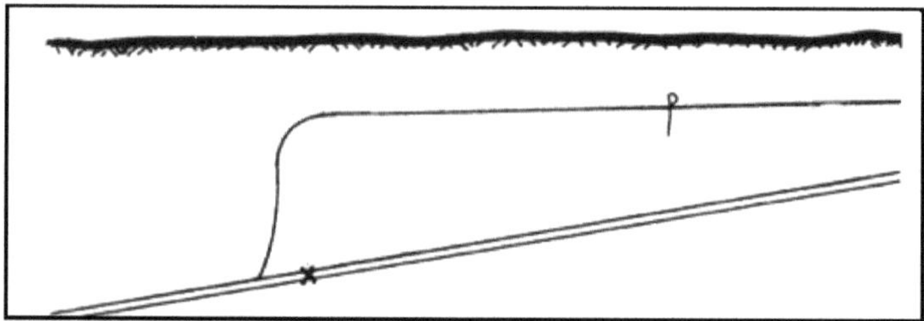

b- Sloping ground

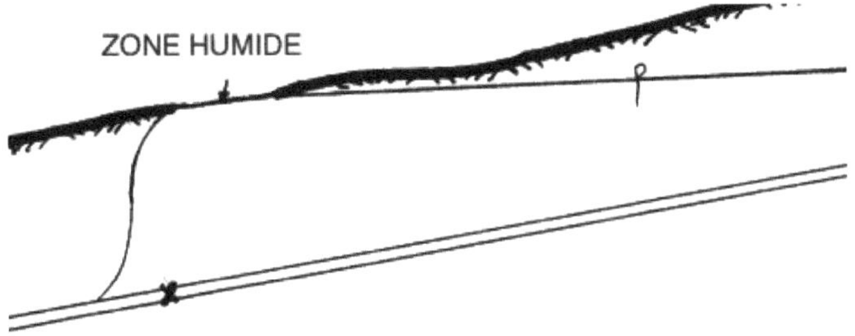

c- Water diffusion pattern

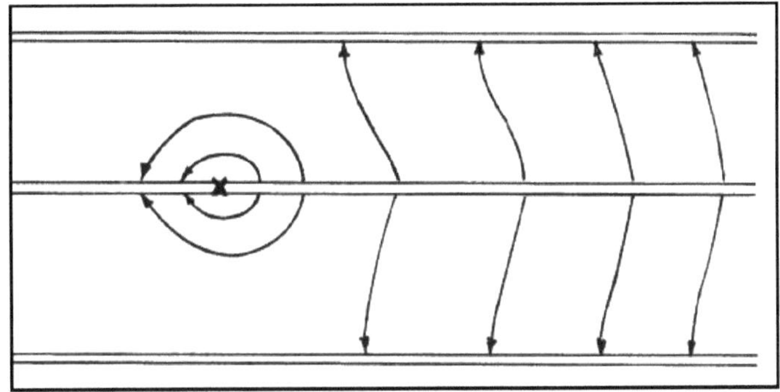

Figure 22: Trapezoidal channel and definition of terms.

As a result, the water table will draw down at approximately one-third of the anticipated rate. If the design has foreseen very rapid drawdowns and the hydraulic conductivity of the soil is very high, the problem will be hardly noticeable. For sloping ground, the problem will have completely different consequences. If upstream water tables are higher than the level in the drain at the problem point, they will drain in proportion to the existing gradient and continuously feed the drain. The flow thus produced will back up when it encounters the problem point and will have to diffuse into the ground to reach the downstream drain and other surrounding drains. As the drain is always fed by upstream water tables, the ground in the vicinity of the break will be continuously very wet, and we'll have the impression of being in the presence of a spring. In some cases, we may even see a trickle of water emerge from the surface. On the other hand, we will have the impression that drainage is functioning more or less normally towards the upstream end of the drain.

The problem is best identified by observing the water table profile transversely and longitudinally to the drain, using piezometers or observation wells. The profile of the water table longitudinally to the drain (preferably within a few centimetres of it) will be almost horizontal upstream of the point of obstruction, and will show an abrupt drop in water level within a few metres downstream (Figure 23). The cross-sectional profile will show a water table as if the drain did not exist. Water pressure in the drain will always be equal to or higher than the surrounding water table. Driving a rod into the drain will cause the water level to rise in a hole drilled above the drain, rather than lowering it.

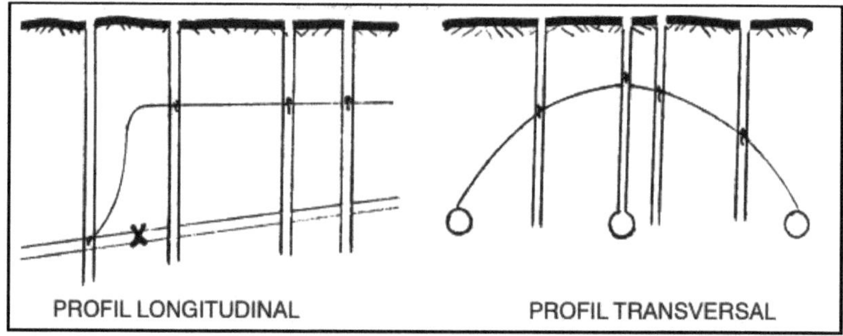

Figure 23: Groundwater profile for a broken, crushed or blocked drain.

When a broken, crushed or blocked drain has been detected, the exact location of the obstruction is determined as follows:

a) Identify the two consecutive piezometers or wells drilled along the drain where a significant drop in water level is observed,

b) dig a well halfway between the two wells showing a significant drop in water level,

c) If the water level stabilizes at the same level as the upstream wells, the problem location is downstream of this well. If the water level stabilizes near the drain level, the problem location is upstream of this well,

d) repeat steps b) and c), halving the distance until the two wells are less than four metres apart,

e) dig between the two shafts and you'll discover the pot aux roses. Don't be surprised if you find yourself working in a lake of water.

This approach only works if the water table is higher than the drain level. It is possible to locate a broken or blocked drain even with a water table 20 to 30 cm above the drains.

For those who find this procedure a little time-consuming, you can replace steps b) to d) by digging a series of wells close together between the two wells showing a significant drop in water level.

VI.3.2. Drain filled with sediment

A drain partially filled with sediment has a free cross-section and reduced flow capacity over the entire affected portion of the system. The drain (lateral or collector) cannot carry all the water that the water table could supply. As a result, the drain will flow like a loaded drain, giving the illusion of a drain installed at shallow depth below the water table. The cross-sectional profile of the water table will show a slight curvature (Figure 24), and the pressure in the drain will be greater than the diameter of the drain, yet less than the hydraulic head of a piezometer adjacent to the drain.

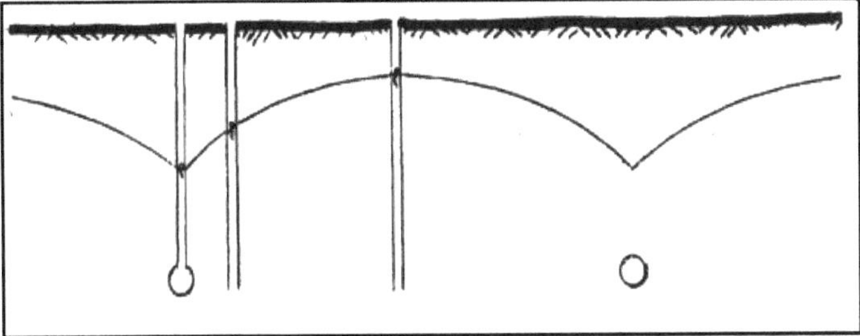

Figure 24: Water table profile for drains partially filled with sediment.

When the water table is very low, the cross-sectional profile of the water table will give the illusion of a normally functioning drain. The system flow rate as a function of water table height will be as shown in Figure 25.

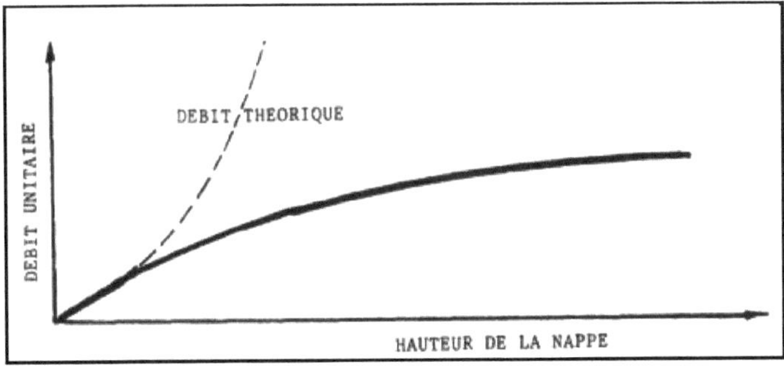

Figure 25: Unit flow of a drain partially filled with sediment.

VI.3.3. Drain clogged around its perimeter

A drain that is clogged all around (external clogging) offers very high resistance to water entering the drain. As a result, most of the available hydraulic head will be used to force water into the drain, and unit flow will be greatly reduced (figure 26).

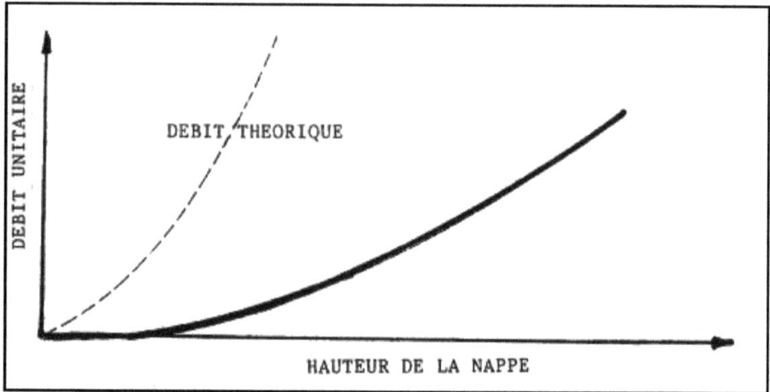

Figure 26: Unit flow rate of a drain clogged around its perimeter.

The cross-sectional profile of the water table will be almost horizontal, with a very high hydraulic head near the drain but very low water pressure in the drain (figure 27).

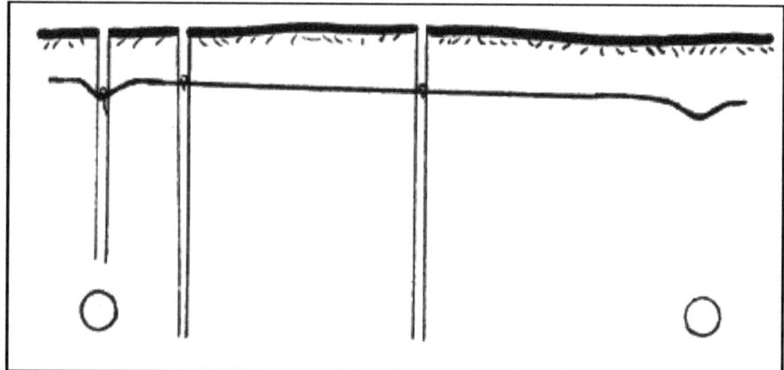

Figure 27: Water table profile of a drain clogged around its perimeter

A clogged drain is difficult to correct. Installing a new drainage system is almost the only solution, and should only be considered once the cause of the clogging has been identified. The main causes are breakage of the soil structure during installation or clogging of the filter around the drain.

VI.3.4. Drain installed in an almost impermeable trench

This is similar to the case of a drain clogged around its perimeter, except that it will be very difficult to identify a hydraulic head above the drain by means of an observation well or piezometer.

This case is mainly encountered in sensitive soils (mainly clays), where installation is carried out when the water table is too high. The vibrations of the machinery destroy the entire soil structure around the trench.

VI.3.5. Drain installed in an impermeable or low hydraulic conductivity horizon

This situation takes two forms:

-the drain has been backfilled with more permeable surface soil, or fracturing of the soil by the paving machine has made it as permeable as the soil in the upper horizon,

- the drain was backfilled with soil from the low-permeability horizon, restoring the soil's low hydraulic conductivity.

In the first case, the cross-sectional profile of the water table is almost horizontal, with a very low hydraulic head in the trench. The unit flow rate (Figure 28) and the water table drawdown correspond to those of a drain installed at the interface of a low hydraulic conductivity horizon. This case will generally cause few problems if the low-permeability horizon is more than 80 cm deep.

As for the second case, it will present the same symptoms as a clogged drain around its perimeter. This case could be avoided if the soil profile was properly identified when the drainage plan was drawn up.

Figure 28: Unit flow rate of a drain installed in a low-permeability horizon, but where the trench is permeable.

VI.3.6. Depressions

Depressions cause major drainage problems. Because of their topographical situation, they are a prime location for runoff to accumulate. What's more, hypodermic runoff can contribute to feeding the depression even when there is no runoff. Depressions are places that remain wet for long periods. This makes plant growth difficult, and machine traffic delayed or problematic. As a result of tillage in wet conditions, depressions often have an indurated layer beneath the plough layer, which considerably slows down the percolation of water towards the water table. A depression is characterized by a water level in the depression that is higher than that of the surrounding water table.

Correcting depression problems is not always easy. If the depression has no indurated layer, it can simply be filled with soil. If the depression does have an indurated layer, the problem simply cannot be solved by filling the depression alone, as the hypodermic flow will continue to feed it. The indurated layer must also be broken up, and this is no easy task.

VI.3.7. Presence of a compact plough layer

A compacted tillage layer will reduce the soil's capacity to infiltrate water, slowing down the rate at which the soil re-dries after a rain. As a result, the ploughed layer will remain wet for a long time after a rain. A layer of water will tend to appear very quickly on the surface of the soil after the onset of precipitation. Observation wells will show a low water table (figure 29) even if the ploughing layer appears saturated, and will sometimes show a perched water table if a ploughing sole is present (figure 30).

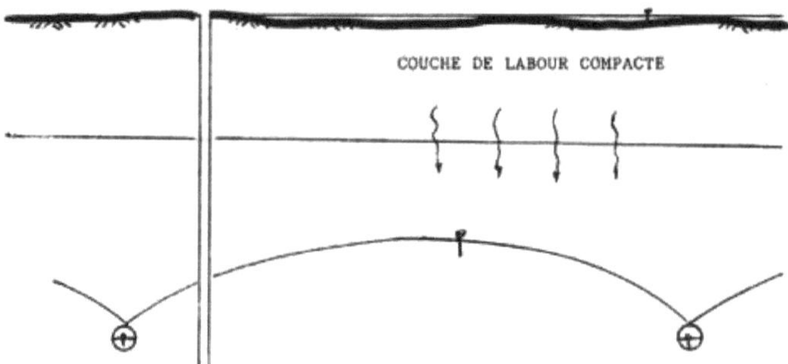

Figure 29: Effect of a compact plough layer on the water table

This situation will reduce the amount of water that can infiltrate to moisten the profile in the dry season (accentuating water deficit problems) and make the plant uncomfortable when it rains. The plant may also suffer from a lack of oxygenation. In such a situation, yields may suffer through no fault of the underground drainage system. This situation is mainly caused by soil compaction and poor soil and crop management. It is often associated with a reduction in organic matter content. The solution is agronomic: better soil and crop management, including crop rotation. Soil decompaction is generally only a short-term solution.

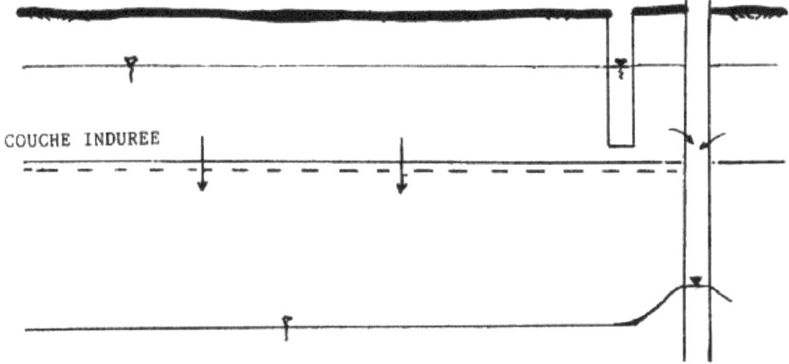

Figure 30: Effect of a plough sole or indurated layer

VI.3.8 Presence of an indurated horizon or a plough sole

The presence of an indurated horizon or a ploughed footing will reduce water percolation towards the water table, creating a perched water table in the upper horizons. The rate at which this perched water table is lowered will depend solely on the rate at which water percolates through the indurated horizon or ploughing footing, and not on the spacing between drains, unless the deep water table reaches the indurated horizon or ploughing footing.

The identification of an indurated horizon or a ploughed footing can be easily made when the soil surface remains moist after precipitation and the perched water table is present. A cross-section of the soil profile will easily show seepage at the interface of the sole or indurated horizon. The use of a deep well (1 - 1.5 m) and a well drilled into the surface horizon will show the characteristic behavior shown in figure 30 following a heavy rainfall. A difference in water level between the shallow and deep wells is a sure sign of the presence of a perched water table and an indurated horizon or a ploughing footing.

The problem can be corrected by subsoiling if the problem horizon is shallow. The solution is not necessarily permanent and the problem may recur.

VI.3.9 Drain installed too shallowly

The main influence of drain depth is on machine traffic conditions. It is recognized that the water table must be at a minimum depth of 50 to 60 cm to provide sufficient bearing capacity for machine traffic. Thus, an underground drainage system designed to lower the water table by 30 cm/d when the water table reaches the ground surface will take 2.8 days to lower the water table from the ground surface to a depth of 60 cm if the system is designed and installed for drains at a depth of one metre. The same system designed and installed for drains 75 cm deep will take 6.7 days. A system designed for drains one metre deep but installed at a depth of 75 cm will take 10 days. This farmer will feel like he's waiting forever to get into his field. The calculations were made assuming a water thickness of 5 cm in the drains and an equivalent drainage depth of one meter.

As a result, it is very difficult to draw down the last 20 cm of water table above the drain due to the low hydraulic gradient. Measuring the depth of the water table and the drains will quickly reveal the problem of insufficiently deep drains. The problem can only be corrected by reinstalling a new drainage system at a suitable depth.

VI.3.10 Too wide a spacing between drains

Too wide a spacing between drains results in a slow lowering of the water table and a regular presence of the water table at less than 60 cm from the soil surface. These high water tables considerably hinder machine traffic and cultivation work. Observing the water table in a well halfway between two drains will show a very slow drawdown.

VI.3.11. Frozen soil

Frozen ground with a high water table can give the illusion that the drainage system is malfunctioning. As the ground thaws, the problem should disappear within a few days. Frozen ground is easily identified with a probe.

VI.3.12. Undersized manifold

An undersized collector will cause it to flow under load as the water table approaches the ground surface. A collector flowing under load means that the gradient available to draw down the water table is reduced, and that the drawdown of the water table is slower than the design specifications (Figure 30). As the effect of headflow is transmitted from downstream to upstream, the areas furthest upstream will be the most affected.

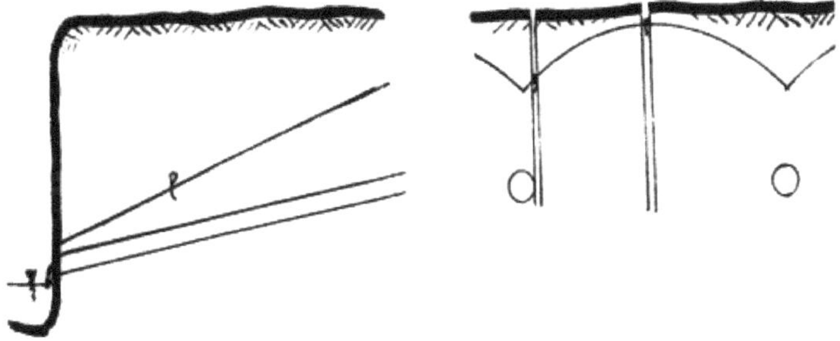

Figure 31: Influence of an undersized collector.

In a chronically undersized system, areas near the outlet will suffer little from headflow, as the potential gradient remains close to design conditions; the water table will draw down almost normally. For upstream sectors, the drawdown may be almost nil if the hydraulic head in the collector reaches the ground surface. Drawdown will only really begin once downstream areas have drained somewhat. An undersized collector delays drainage of upstream areas when the water table rises close to the ground surface. This effect is only really noticeable in very long collectors. The maximum flow that a collector can deliver would occur when the hydraulic gradient corresponds to the difference in height between the ground level at the furthest point from the collector and the collector outlet. This maximum flow should be several times the design flow. Apart from design error, the main causes of undersizing are underestimation of hydraulic conductivity or equivalent drainage depth.

VI.3.13. Shallow watercourse

The effect of a watercourse where drains exit below the water level is to reduce the hydraulic gradient. Thus, the hydraulic gradient, instead of corresponding to the drain slope, corresponds to the difference in elevation with the water level in the watercourse (Figure 32). When the hydraulic gradient is significantly reduced, the hydraulic gradient required to draw down the water table is lower than anticipated at the design stage. This slows down the lowering of the water table, as if the drains were installed at a shallower depth. This effect is only visible when the watercourse is full over a long period.

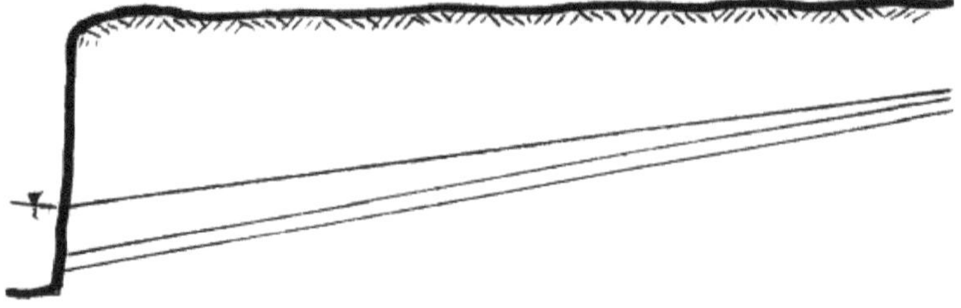

Figure 32: Influence of an undersized manifold

VI.4 Problem identification methodology

Now that we know the various problems and the symptoms associated with them, we can define an effective methodology for identifying problems in an underground drainage system.

The first step is to obtain from the farmer the best possible description of the problem (manifestations, locations, frequencies, etc.) and the conditions under which it was observed. The engineer must also collect the drainage plan and reports, which should include all soil studies (soil profile description, hydraulic conductivity, granulometry, thickness of different horizons and depth of permeable soil) and design criteria. This first step provides a subjective assessment of the problem(s) and whether it is localized or generalized. A problem is considered generalized if it affects an entire plot or drainage system, and localized if it affects only part of the system.

The second step is to draw up an observation plan to try and objectively assess the operation of the drainage system under the conditions in which the problem occurs. Drains must be located and a series of observation wells dug at strategic points to determine the shape of the water table, the hydraulic load in the vicinity and the pressure in the drain. Wells and drains must be levelled. Locating drains is probably the most time-consuming operation. This step can be minimized if you suspect the presence of an indurated layer, a low-permeability plough horizon or a depression problem.

The third step is to measure the behavior of water levels in the wells and the flow rate of the collectors (if possible) when the problem occurs. The farmer can play an active role in this stage. A visit to the site at this time is very useful. This stage generally takes place in autumn or spring, as the water table is generally high and problems are more visible at this time. This stage is essential, otherwise it's all speculation.

The fourth step consists of analyzing the behavior of water levels to assess the extent of the problem, the level of efficiency or inefficiency of the system, and to reveal numerous symptoms which, when compared with the symptoms in section VI.4, will point to possible causes of the problem(s). If the problem corresponds to one of the simple, straightforward cases presented in section VI.4, problem identification will be rapid. On the other hand, if the problem is the result of two or more of the cases presented in section VI.4, identification becomes more complex and may require additional observations.

The final step consists, if necessary, in digging out the drain at critical points. This step, carried out without the others, is often disappointing, as it only identifies cases where the drain is filled with sediment. What's more, digging up drains that are below water level doesn't reveal much. The most frequent problems I encountered were crushed, broken or blocked drains, drains filled with sediment, the presence of indurated layers, a low-permeability plough layer, drains installed in a low-permeability horizon, the presence of depressions and drains installed too shallowly. Many of the problems encountered could have been avoided if the soil had been examined sufficiently before installing the underground drainage system.

VI.5. Conclusions

This study presented the main problems likely to be encountered in underground drainage systems, the symptoms associated with them, and a methodology for identifying them. Identifying underground drainage problems and their causes requires an excellent theoretical knowledge of all the processes involved in underground drainage, as well as excellent powers of observation and deduction.

Chapter VII: Machines used

VII.1 Introduction

Various machines can be used to install underground drainage pipe. The most commonly used are :

- The backhoe
- The wheel excavator
- Chain excavators
- The mole-plough

VII.2 Backhoe

Backhoes (figure 33) are generally tractor-mounted, and use a bucket with a maximum cubic capacity of 0.385 m^3 or 0.480 m[3]. They are generally used for small-scale underground drainage work in rocky and uneven terrain. They are mainly used on construction sites to excavate trenches to connect drains to a collector, a collector to another collector, or trenches to install special structures.

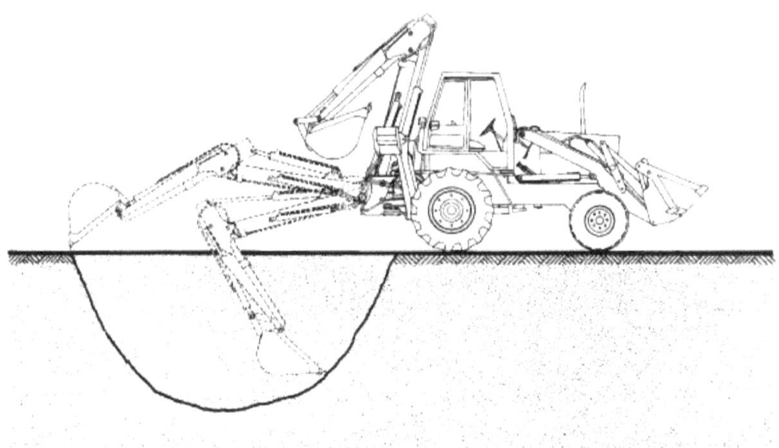

Figure 33: Backhoe (CPVQ, 1976)

VII.3 Wheel excavator

The wheel excavator (figure 34) features a large wheel mounted on a frame at the rear of the machine. The position of this wheel varies independently of the machine to maintain the given slope. Attached to this wheel, buckets feed the excavated soil onto a conveyor which deposits it on one side or the other of the trench. At the rear of the wheel, a caisson prevents the soil from falling back into the trench, and a shoe shapes a furrow in the trench bottom to better seat the drain. The caisson is long enough to keep the trench clean for the drain and filter material.

Wheel excavators used in North America generally dig trenches 55 centimetres wide and up to 1.8 metres deep. The excavator can be mounted on tracks or tires.

Wheeled excavators are rarely used, as they are slower than mole ploughs.

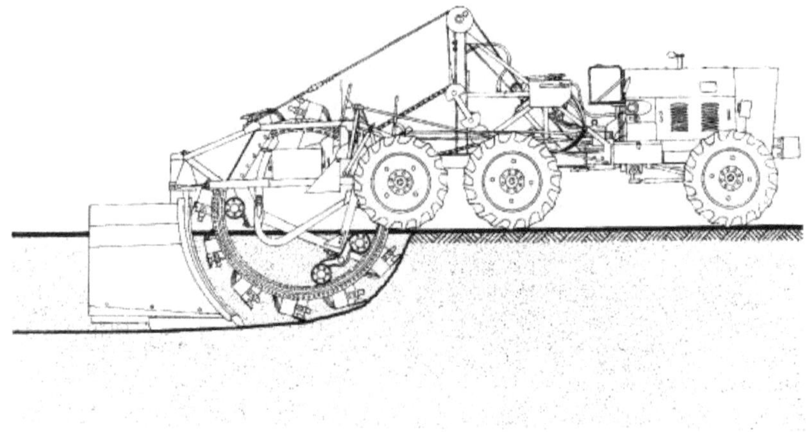

Figure 34: Wheel excavator (CPVQ, 1976)

VII.4 Chain excavator

With chain excavators (Figure 35), excavation is carried out by means of an endless chain fitted with buckets and working vertically or at an angle of 45°. The excavated soil is deposited on either side of the trench. Chain excavators generally make trenches 25 to 40 centimetres wide. Chain excavators cannot work in rocky terrain.

Like the wheel excavator, chain excavators are becoming less and less common, and are mainly reserved for draining organic soils or sensitive clays.

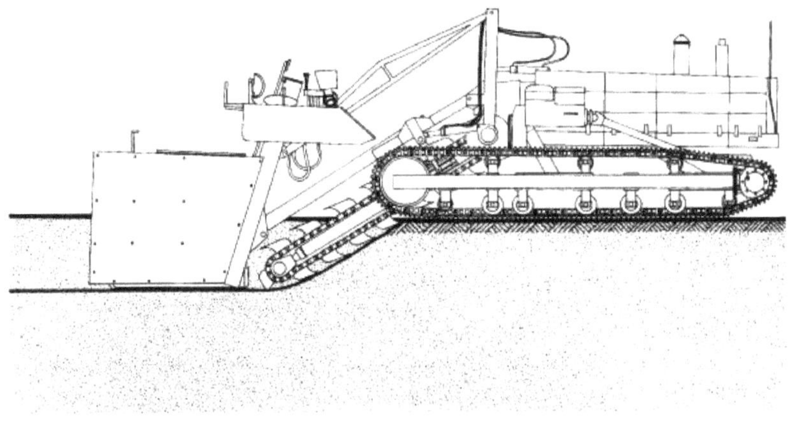

Figure 35: Chain excavator (CPVQ, 1976)

VII.5. Mole plough

The mole plough consists (figures 36 and 37) of a subsoiler coulter connected to a crawler tractor (or dozer) by means of hydraulic arms and cylinders. The subsoiler coulter is fitted with a cutting plate and a tooth tip. The drain is conveyed into the caisson by means of a chute. At the rear of the caisson, a shoe digs into the ground to accommodate the drain. Cylinders are used to adjust the depth and angle of attack of the subsoiler coulter.

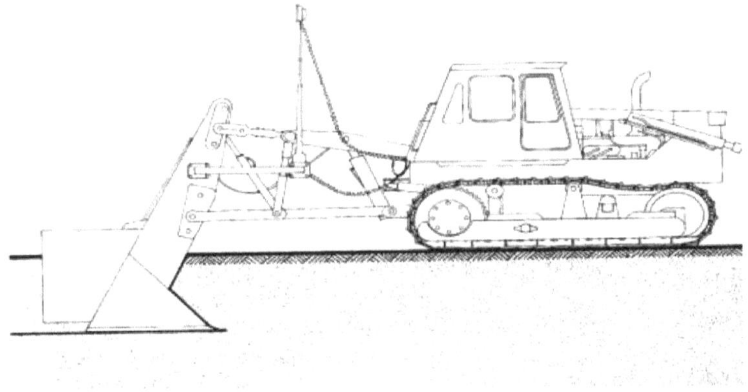

Figure 36: Mole plough (CPVQ, 1976)

VII.6 LASER system

Figure 37 shows a diagram of a laser guidance system for controlling installation depth.

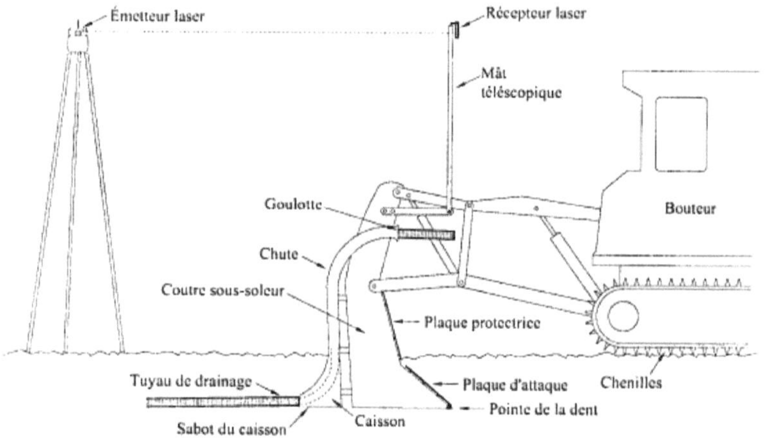

Figure 37: Diagram of a laser guidance system (BNQ, 2005)

Chapter VIII: Layout principles and drainage system modelling approaches

VIII.1 Introduction

A drainage network is made up of drains (ditches or buried pipes), secondary collectors transferring the flows from the drains (ditches or buried pipes), main collectors (ditches or buried pipes) transferring the cumulative flows from one or more secondary collectors to an outfall (wadi, sebkhat etc.).

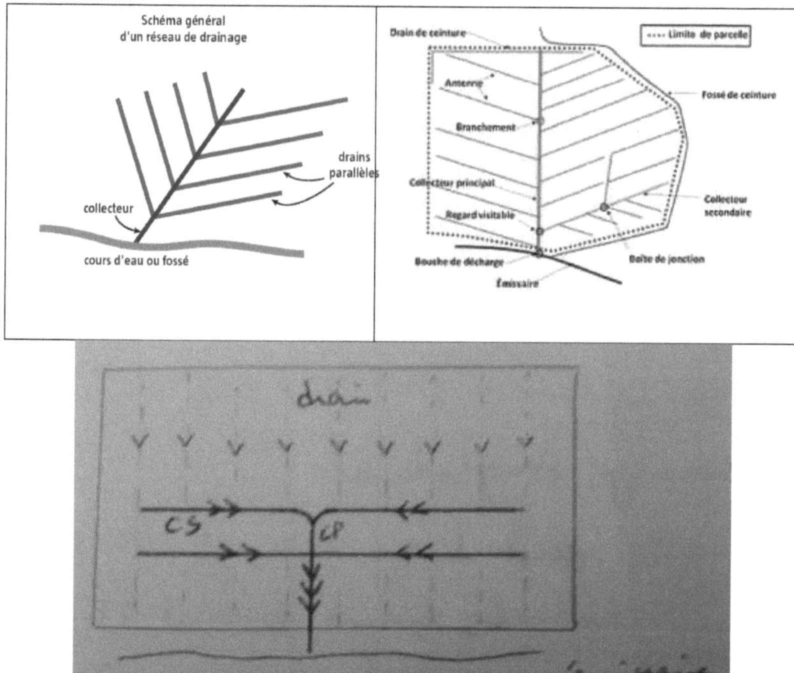

There are two principles (methods) for tracing a drainage network.

VIII.2 Principle of cross-sectional layout

In this principle, the drains are oblique to the contour lines (CN) and the secondary collectors (CS) are perpendicular to the CN.

- In this case, the collector had to be placed in the Talweg, and the velocity in the collector had to be higher than that in the drain ($V_{(col)}$> V_{drain}) to enable self-cleaning of the network.)

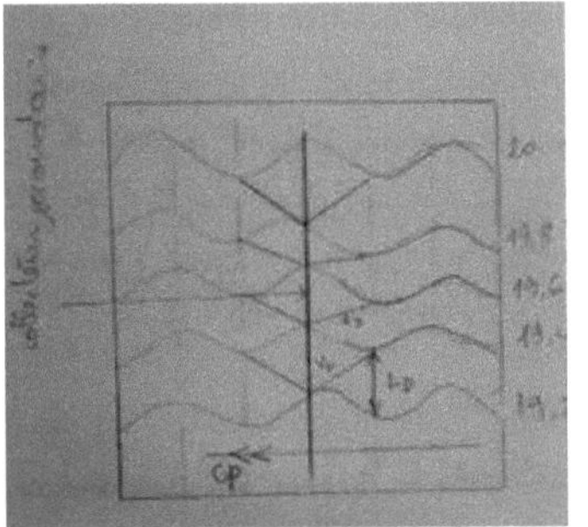

VIII.3. Principle of longitudinal alignment :

Drains are perpendicular to the NC and collectors are oblique to the NC. If possible, drains should be placed at talweg level to facilitate self-cleaning.

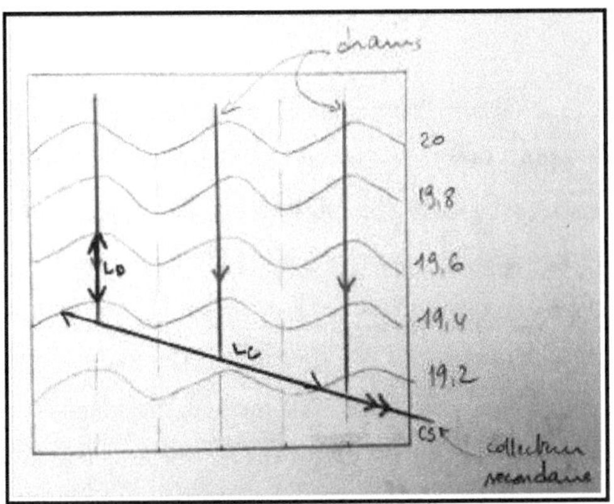

V_{neck} must $< V_{drain}$ to allow the network to self-clean.

The slope that ensures normal flow in drains is generally 2‰.

VIII.4. Drainage system modelling approaches

There are a number of different approaches to the design of agricultural subsurface drainage systems, differing from one another in terms of their purpose and the way in which processes are described. The development of steady-state flow equations, to facilitate the design of parallel drainage systems in a field, began in the middle of the twentieth century. A steady-state drainage equation for homogeneous soils with two distinct layers was presented by Hooghoudt (1940). In this case, the upper layer is highly permeable, while the lower layer has low permeability. Ernst (1962) derived an equation for soils with a low-permeability layer above a high-permeability layer, and a modified version for a case with two permeable layers below the drain level. The approaches of Hooghoudt (1940) and Ernst (1962) were combined into a single equation applicable to all the cases presented (Ernst, 1975). Principled derivations of the Hooghoudt and Ernst equations were presented by van Beers (1976). Other approaches to modelling underground drainage have been proposed, such as the Kirkham equation (Kirkham, 1958) and the Laplace or Boussinesq equations (Bouarfa and Zimmer, 2000). These approaches have been integrated into several conceptual and distributed models. The Kirkham and Hooghoudt equations are used, for example, in the DRAINMOD model (Skaggs, 1980), while the MACRO model (Larsbo and Jarvis, 2003) uses the Hooghoudt equation. In the SWAP model (Kroes et al. 2008), the user can choose between the Hooghoudt equation option and the Ernst equations, depending on the problem under study.

In modelling flows in a variably saturated regime, drains are generally simulated in a simpler way than in steady-state approaches, by virtue of the fact that soil layers and pressure distributions are taken into account by the model. Fipps et al (1986) tested four different methods of representing subsurface drains in a numerical finite element solution of Richard's equation in two dimensions: (1) hole (opening) approach as drain model, (2) nodal point approach with imposed or specified flow, (3) imposed pressure approach and (4) resistance adjustment approach. In this last method, the conductivity at the drains is adjusted by a factor determined by the ratio between the effective drain radius and the size of the elements surrounding the drain node. The problem with methods (1) and (3) is that a large number of nodes are required in the vicinity of the drain for more realistic results, whereas methods (2) and (4) are implemented with coarse grids. In the HYDRUS model (Simunek et *al.* 2006), drains can be simulated by adjusting the hydraulic conductivity of the elements surrounding the drain nodes. The drainage system is based on the electrical resistance network or considered as seepage faces around their perimeter. Karvonen (1988) made an in-depth comparison between (a) a two-dimensional finite element approach with a progressively denser grid around the drain, (b) an analytical model and (c) a one-dimensional approximation. He concludes that there is no major benefit to be gained from complicated modeling of drain geometry compared with simpler approaches, due to the uncertainty of available field data. For three-dimensional grids where there are a number of drains in the domain, the extremely fine discretization required to represent small openings (drains) is computationally time-consuming (Fipps et al. 1986). To obtain a solution to the flow problem, a Dirichlet (imposed hydraulic pressure) or Neumann (imposed flow) boundary condition is generally applied to the nodes representing the drains. One of these two conditions can be integrated into the CATHY model (Camporese et al. 2010) to simulate the underground drainage network. The use of the Dirichlet condition implies that the potential at the drains remains constant throughout the simulation and the flow in the drains is not affected by factors such as drain cross-section (MacQuarrie and Sudicky, 1996). In order to simulate the drainage system in the three-dimensional mesh of the porous matrix, where groundwater flow is governed by Richard's equation, the following three approaches can be distinguished. Firstly, a one-dimensional drain flow equation is introduced. This is the case with the HydroGeoSphere model (Brunner and Simmons, 2012; MacQuarrie and Sudicky, 1996; Therrien et *al.* 2007), where it is based on the continuity equation for open channel flow. Then, the outflow from the drainage system depends on the height of the water table above the drain level and an imposed time constant that is calculated according to the linear reservoir concept. This concerns the MIKE-SHE model (DHI, 2007). And finally, we can resort to the equivalent representation of buried drains in the soil profile using an anisotropic homogeneous porous medium (Carlier et al. 2007). This latter approach, compared with others that represent underground drains using the SWMS 3D code (Simunek et al. 1995), had produced satisfactory results in terms of flow rates and mean water table elevation. Of all the above hydrological models, it should be noted that, on the one hand, very few, if any, studies using the CATHY model have considered the modelling of agricultural subsurface drainage. On the other hand, the DRAINMOD model is widely used worldwide to describe the hydrology of poorly drained or artificially drained soils. This explains the choice of these two models for one or other of the studies in this research.

VIII.4.1. Description of the DRAINMOD model

Widely used throughout the world due to its ease of use, DRAINMOD (Skaggs, 1978) is a physics-based 1-D model composed of two modules: the first deals with hydrology (water balance at the soil surface and in the porous medium) and the second with nitrogen transport (this module is not the subject of this study).

DRAINMOD has been developed for use at field scale to describe the hydrology of poorly drained or artificially drained soils. The model can also be used to simulate the hydrology of non-drained soils containing wetlands. Hydrological variables (infiltration, subsurface drainage, surface runoff, evapotranspiration, vertical percolation and hypodermic flow, water table depth, etc.) are predicted and output summaries are available on a daily, monthly or annual basis, depending on the user's option.

References and bibliographies

BNQ, 2005. Agricultural subsurface drainage service -- Quality criteria. BNQ 3624--540/2005. Bureau de normalisation du Québec, Quebec City.

Bouarfa S & Zimmer D (2000) Water-table shapes and drain flow rates in shallow drainage systems. Journal of Hydrology 235(3-4):264-275.

Brunner P & Simmons CT (2012) HydroGeoSphere: A Fully Integrated, Physically Based Hydrological Model. Ground Water 50(2): 170-176.

Camporese M, Paniconi C, Putti M & Orlandini S (2010) Surface-subsurface flow modeling with path-based runoff routing, boundary condition-based coupling, and assimilation of multisource observation data. Water Resources Research 46(2):W02512.

Carlier JP, Kao C & Ginzburg I (2007) Field-scale modeling of subsurface tile-drained soils using an equivalent-medium approach. Journal of Hydrology 341(1-2):105-115.

CPVQ, 1976. Underground drainage, general information. Conseil des productions végétales du Québec.

CPVQ. 1989. Underground drainage -- Specifications. Conseil des productions végétales du Québec, Quebec City. Agdex 555.

CRAAQ. 2005. Guide de référence technique en drainage souterrain et travaux accessoires. Center de référence en agriculture et agroalimentaire du Québec. Quebec City.

Colwell HTM (1978) The economics of increasing crop productivity in Ontario and Quebec by tile drainage installation. Canadian Farm Economics 13(3):1-7.

DHI (2007) MIKE-SHE User Manual, Volume 2: Reference Guide. DHI Water and Environment, Danish Hydraulic Institute, Denmark.

Ernst LF (1962) Grondwaterstromingen in de verzadigde zone en hun berekening bij aanwezigheid vanhorizontale evenwijdige open leidingen. Versl. Landbouwk. Onderz. 67.15. PUDOC, Wageningen. 189 p.

Ernst LF (1975) Formulae for groundwater flow in areas with subirrigation by means of open conduits with a raised water level, Misc. Reprints 178, Institute for Land and Water Management Research, Wageningen, The Netherlands, 32 p.

Fipps G, Skaggs RW & Nieber JL (1986) Drains as a boundary condition in finite elements. Water Resources Research 22(11):1613-1621.

Gallichand J. and R. Lagacé. 1987a. Modeling sediment movement into perforated subsurface drains. Transaction of ASAE 30(1):119--124.

Gallichand J. and R. Lagacé. 1987b. Verification of a model predicting sediment level in subsurface drains. Transaction of ASAE 30(6):1648--1652.

Gallichand J., R. Lagacé and M. Cailler. 1989. Thin--section study of soil near corrugated subsurface drain perforations. Geoderma 43:337--347.

Gallichand J. and R. Lagacé. 1991. Effect of perforation dimensions and hydraulic gradient on sedimentation in subsurface drains. Can. Agr. Eng. 33(1):17--25.

Hooghoudt S (1940) Hooghoudt's theory of drainage. Technical report, Institut voor Cultuurtechnik en Waterhuishouding, The Netherlands (in Dutch).

Karvonen T (1988) A model for predicting the effect of drainage on soil moisture, soil temperature and crop yield, Doctoral dissertation, Helsinki University of Technology, 215 p.

Kroes, Van Dam, Groenendijk P, Hendriks RFA & Jacobs CMJ (2008) SWAP version 3.2 Theory description and user manual, Alterrareport, Wageningen, 262 p.

Kirkham D (1958) Seepage of steady rainfall through soil into drains. Transactions of American Geophysical Union 39(5):892-908.

Lagacé R. and R. W. Skaggs. 1985a. Prediction of drain clogging by the aggregate analysis method. In Underground drainage research. 12th Colloque de génie rural, Université Laval, Québec. GR--H--85--01:19--39.

Lagacé R. and R. W. Skaggs. 1985b. Drain clogging: Laboratory experiments. In: La recherche en drainage souterrain. 12th Colloque de génie rural, Université Laval, Québec. GR--H--85--01:41--68.

Lagacé R. and R. W. Skaggs. 1988. Predicting of drain silting by soil aggregate size analysis. Transaction of ASAE 30(6):1648--1652.

Lagacé R., R. W. Skaggs and J. Gallichand. 1987. Prediction of drain sedimentation. Proceeding 5th National Drainage Symposium. ASAE Publication 07--87:354-361.

Larsbo M & Jarvis N (2003) MACRO 5.0. A model of water flow and solute transport in macroporous soil. Technical description, Swedish University of Agricultural Sciences, 40 p.

MacQuarrie KTB & Sudicky EA (1996) On the incorporation of drains into three-dimensional variably saturated groundwater flow models. Water Resources Research 32(2):477-482.

Šimůnek J, van Genuchten M & Šejna M (2006) The HYDRUS software package for simulating the two- and three-dimensional movement of water, heat and multiple solutes in variablysaturated media, Technical manual, 241 p.

Skaggs RW (1978) A water management model for shallow water table soils. Technical Report No. 134. Water Resources Research Institute, North Carolina State University, Raleigh, N.C.

Van Beers WFJ (1976) Computing drain spacings, Bull. 15, ILRI, Wageningen, The Netherlands.

Printed by Books on Demand GmbH, Norderstedt / Germany